UNITED NATIONS CONFERENCE ON TRADE AND DEVELOPMENT
Geneva

International Accounting and Reporting Issues

1995 Review

Environmental Accounting

Report by the secretariat of the
United Nations Conference on Trade and Development

UNITED NATIONS
New York and Geneva, 1996

NOTE

Symbols of United Nations documents are composed of capital letters combined with figures. Mention of such a symbol indicates a reference to a United Nations document.

*
*   *

The designations employed and the presentation of the material in this publication do not imply the expression of any opinion whatsoever on the part of the Secretariat of the United Nations concerning the legal status of any country, territory, city or area, or of its authorities, or concerning the delimitation of its frontiers or boundaries.

*
*   *

Material in this publication may be freely quoted or reprinted, but acknowledgement is requested, together with a reference to the document symbol. A copy of the publication containing the quotation or reprint should be sent to the UNCTAD secretariat.

UNCTAD/DTCI/25

UNITED NATIONS PUBLICATION
Sales No. E.95.II.A.11
ISBN 92-1-104452-9

PREFACE

One of the most common questions asked about environmental accounting is, very simply, "What is it all about?". This publication is not an extensive text on all aspects of environmental accounting but it does describe some of its different aspects from both a theoretical and practical perspective. The most straightforward topic in environmental accounting is disclosure of relevant information by transnational corporations. Environmental accounting becomes more sophisticated when it addresses such issues as the valuation of natural resources consumed during the production process, or the value of the waste disposal service that the environment provides, or the recognition and valuation of environmental liabilities.

Environmental accounting has been discussed extensively since 1990 by the Intergovernmental Working Group of Experts on International Standards of Accounting and Reporting (ISAR) and by many other groups such as: various units of the United Nations system; international and national accounting and reporting standard setting bodies; regional and national professional accounting organizations; and public and private sector corporations. Since ISAR first considered this issue it has frequently returned to the topic with new ideas and perspectives. To date, the most comprehensive summary of the Group's recommendations on environmental accounting can be found in the publication *Conclusions on Accounting and Reporting by Transnational Corporations* (UNCTAD/DTCI/1).

This year (1995) the thirteenth session of ISAR was devoted exclusively to the subject of environmental accounting. The Chairman of the session, Mr. L. Nelson de Carvalho (Brazil), felt that the high level of participation at the session was indicative of the prominence that environmental issues have on the international agenda.

Four substantive papers were discussed during the thirteenth session and this research forms the basis for the first four chapters of this volume. These papers concern: incentives and disincentives for the adoption of the concept of sustainable development by transnational corporations (Chapter I); environmental performance indicators (Chapter II); a review of national environmental accounting rules and regulations (Chapter III); and disclosures by transnational corporations of environmental matters at the national level in annual reports (Chapter IV). Mr. Robert Gray, University of Dundee (United Kingdom) was present to provide expert advice on various aspects of environmental accounting throughout the session and Mr. Claus Noppeneny, University of St. Gallen (Switzerland), gave specific insight into the research on environmental disclosures at the national level.

Also at the session, two experts in environmental management and accounting matters made oral presentations. The first was Mr. Matteo Bartolomeo, Fondazione Eni Enrico Matteo (Italy), who described an environmental balance sheet that he was instrumental in developing. His presentation is included in Chapter V of this volume. Then, Mr. Peter Wilson of the European Commission explained its Eco-Management and Auditing Scheme (see Chapter VI). The aim of these presentations was to describe some current usages of environmental accounting concepts and to demonstrate that, in some instances, the theories have been converted into practical applications.

Additionally, UNCTAD's recent publication, *Accounting for Sustainable Forestry Management: A case study* (UNCTAD/DTCI/4) was presented for discussion by the principal researcher, Mr. Daniel Rubenstein of the Office of the Auditor General of Canada. Supplemental information about the publication and comments from Mr. Rubenstein and the experts at the ISAR meeting are contained in Chapter VII.

Each of the substantive research papers and presentations shed very valuable light upon their subject matter. The previously mentioned "incentives" paper (Chapter I) concluded, among other matters, that very few companies, even leading "environmental" companies, have yet come to terms with the nature of sustainability and its implications for current business practices. Traditional accounting places a considerable constraint upon environmental managers' ability to redress the situation. The environmental performance indicators paper (Chapter II) questions the potential for the incorporation of business and environmental performance indicators. It was noted that "...there are limited prospects for developing environmental reports in monetary terms for managerial use. Still, the approach is in high esteem in academic and political institutions". However, a technique known as Life Cycle Analysis which traces all the environmental relationships when input components enter the production process until they reach their ultimate disposition, in other words "cradle" to "grave", holds considerable hope as a means of comprehensive environmental reporting. This possibility was also supported in the "incentives" paper.

The paper on disclosure of environmental matters at the national level (Chapter IV) found that there are certainly instances in which TNCs do not report environmental matters consistently throughout the world in the various host countries in which they operate. However, there is a possibility that some TNCs may be in the process of developing comprehensive environmental reporting systems which will be introduced first in the home countries and will then be implemented subsequently on a worldwide basis. Certainly, the lack of environmental accounting laws and regulations in many countries, based upon the survey performed (see Chapter III), could be a significant contributory factor, but increased public concern may contribute to greater real progress in environmental disclosures by TNCs in the host countries than stringent legal requirements.

The experts concluded the session with a recommendation for UNCTAD to conduct a series of workshops on environmental accounting. International experts and researchers should be invited to participate in the workshops to develop guidelines on environmental accounting which would be useful for national standard setters. It is anticipated that some of the best ideas from around the world would be drawn together to form the basis for these guidelines. The first workshop was held during December 1995.

The encouragement of harmonization of international accounting and reporting standards for transnational corporations is the principal objective of ISAR and this programme is one of the mandates of the United Nations Conference on Trade and Development. The Accounting Section within UNCTAD's Division on Transnational Corporations and Investment (DTCI) accomplishes this mandate through research, publications, and support to the annual and ad hoc sessions of ISAR which discuss the most current and important issues in the field of accountancy facing countries. The Accounting Section also provides technical assistance to member States on accounting standards, accounting and tax laws and investment regulations.

The four research reports were prepared for this volume by Robert Gray and Jan Bebbington (Chapter I), Sören Bergström, Bino Catasus, Maths Lundgren and Hams Rämö (Chapter II), Robert Smith (Chapter III) and Claus Noppeney (Chapter IV).

UNCTAD expresses its appreciation to all persons who participated in the session. Also, it is grateful to the Government of Sweden for its financial assistance for the research projects on environmental accounting which UNCTAD has undertaken.

Rubens Ricupero
Secretary-General
United Nations Conference on Trade and Development

Geneva, June 1996

CONTENTS

CHAPTER III

REVIEW OF NATIONAL ENVIRONMENTAL ACCOUNTING LAWS AND REGULATIONS

CHAPTER IV

DISCLOSURE BY TRANSNATIONAL CORPORATIONS OF ENVIRONMENTAL MATTERS AT THE NATIONAL LEVEL

CHAPTER V

A PRACTICAL APPROACH TO INTEGRATED ENVIRONMENTAL ACCOUNTING: AGIP PETROLI CASE STUDY

CHAPTER VI

THE EUROPEAN COMMUNITY'S ECO-MANAGEMENT AND AUDIT SCHEME

CHAPTER VII

SUSTAINABLE FORESTRY OPERATIONS AND ACCOUNTANCY

List of figures

List of tables

Chapter I

INCENTIVES AND DISINCENTIVES FOR THE ADOPTION OF SUSTAINABLE DEVELOPMENT BY TRANSNATIONAL CORPORATION

Report by the UNCTAD secretariat

Summary

Sustainable development has become one of the most widely discussed topics in transnational corporations (TNCs), governmental bodies and academic institutions. However, sustainable development and commercial activity are often considered to be mutually exclusive. This report explores transnational corporations' reaction to this concept based on a survey about aspects of the nature of sustainability, what it means for the enterprise and its business activities, and the role that accounting can, and could, play in the pursuit of sustainability. The report also describes the factors behind the hypothesis of sustainability. In many instances, enterprise executives are not fully aware of the implications of environmental sustainability in the broader context. While a number of TNCs may wish to operate in a more environmentally friendly manner, financial pressure, contemporary markets and the traditional accounting model based upon historical costs are some of the factors which inhibit any substantial changes in TNCs' behaviour. The lack of understanding and formal guidance contribute significantly to the disparity. Furthermore, business-related research needs to be done in this area. Greater support from society as a whole can contribute to sustainable commercial activities.

I. BACKGROUND TO THE STUDY

A. Introduction

Perhaps the most important positive result to arise from the global environmental debate following the 1987 Brundtland Report[1] has been the almost universal acceptance of the goal of sustainability and sustainable development[2]. There has been an explosion of research into different aspects of sustainability and, at a more immediate level, nation States are producing sustainability plans, the European Union has set out its Fifth Action Programme entitled "Towards Sustainability"[3] and business, politicians, NGOs and local governments, among others, are all publicly recognizing the applicability of sustainability to their sector and areas of expertise.

What has not emerged, however, is a recognized consensus on what a sustainable global economy might look like, how we might get there and what implications this "path towards sustainability" might have for current ways of life. More especially, under the gloss of rhetoric, it is very unclear what business is doing, should do or, indeed, can do to re-direct economic activity towards more sustainable development.

This is a far from trivial issue. While it is clear that traditional economic activity and environmental degradation are closely linked, it is far from clear what degree of responsibility for this situation should be, or can be, laid on the shoulders of business. Global society has a right to expect business to do that at which it is most accomplished, i.e. to pursue traditional modes of efficiency, to seek market-led innovation and to respond rapidly and successfully to changes in the "playing field" -- changes in markets, prices, incentives, tastes and so on. It is not clear whether business can be expected to provide, on its own initiative, the innovative ways of thinking, the drastic re-design of life-styles, the costly structural re-adjustments and the major redistribution of wealth which are patently essential for a sustainable future. In this context, for example, the International Chamber of Commerce's (ICC) Charter for Sustainable Business[4] does not actually mention sustainability; and the influential analyses provided by, for example, The Business Council for Sustainable Development (BCSD) and The International Institute for Sustainable Development (IISD) offer absolutely no diagnosis of what sustainability actually might mean for business nor any assessment of how it might be achieved. What is offered in these documents, and correctly so from a business perspective, is a programme of actions which business can take which will be broadly sound economically and reduce the environmental impact of business activity. These actions can be effectively categorized as "pollution prevention pays" (PPP). The Environmental Management and Audit Systems (EMAS)[5] is a response to "green" consumers and the development of supplier chain audits. They have achieved considerable improvements in leading businesses. These steps must be further developed if environmental protection is to be effective. Indeed, such steps may very well be essential pre-requisites for any approach towards sustainability. However, such steps do not necessarily relate directly to sustainability and there is no reason why they might assure sustainability[6].

If we are to embark on the path towards sustainability and business is to play a major role in that journey, EMAS and related initiatives open the gate to that path. They represent little or no progress on the path itself. The reason for this lies in the nature of sustainability itself.

B. The nature of sustainability

Sustainability is an exceptionally difficult concept to define precisely[7]. The general definition of sustainability which is most widely accepted is that provided by the Brundtland Report and refers to development which:

"...meets the needs of the present without compromising the ability of future generations to meet their own needs."[8]

While this proves to be a very good starting point, it is far too general a definition to be practicable. Indeed, its very generality has probably been the reason why the definition is so widely accepted[9]. However, it is proving increasingly possible to define elements of sustainability in a way which is both practicable and recognizably in harmony with the general definition. Two widely accepted and usefully defined elements of sustainability were employed in this study and these are outlined here.

First, sustainability is not only an environmental issue, it is also a social issue. That is, sustainability has essentially two closely related components which one can refer to as "eco-efficiency" and "eco-justice". The first refers to the issues of the physical environment and mankind's use of it, the second to the issues of inter- and intra-generational equity which are essentially social issues[10]. Whilst by far the greatest attention has been paid to the eco-efficiency concerns -- especially so in the business context -- the path towards sustainability must, by definition, involve some advancement in eco-justice issues.

Second, environmental economists have successfully provided a useful and practicable articulation of a number of the essential parameters of eco-efficiency and the inter-generational (but not intra-generational) equity/eco-justice issues[11]. First, we might usefully think of the planet and mankind's interaction with it as involving two kinds of things which we can call "Natural Capital" and "Man-made (or Created) Capital". The natural capital comprises the planet's resources (including eco-systems and waste-sink capacities) whilst man-made capital such as buildings, roads, knowledge, machines and consumer goods, which mankind has "created" from its own efforts employing the resources of the planet. There are two points to note here: (a) in general terms, most of the man-made capital is created at the expense of the natural capital so that as man-made capital rises, natural capital must fall; (b) in general terms, man-made capital is recognised by economic systems (and by accounting systems, see below) in prices whilst natural capital is not recognised. Much of the natural capital will therefore be treated as a free good and its decline will not register in measures of economic success such as GNP or accounting profit.

Furthermore, natural capital can be thought of as comprising two sub-categories: critical natural capital and renewable, replaceable or substitutable (or other) natural capital. The critical natural capital is that essential to the maintenance of life[12] and comprises the integrity of eco-systems, issues such as the maintenance of the ozone layer and protection of species which are close to levels of self-sustenance. Other natural capital is that which can be renewed (e.g. species through breeding or relocation of sensitive eco-systems), repaired (e.g. reclamation of desert or replanting of forests) or substituted or replaced (e.g. use of man-made substitutes for natural resources such as solar panels for fossil fuels).

A sustainable economic system is one which explicitly protects all critical capital, renews some elements of other natural capital and/or employs additions to man-made capital to substitute for depletions in other areas of natural capital. In addition, a sustainable economy would also address the massive inequities between nations -- most obviously the North-South inequality -- in recognition of the intra-generational equity demand of sustainability. It is perfectly clear that no economic system (certainly no "developed" economic system) is currently sustainable under this analysis[13]. The question which then arises is "how large is the gap between current practice and the desired state of sustainability?".

It is probably impossible to answer this question accurately but that may not be important. For example, it is possible to undertake calculations to estimate the sort of expenditure that an economic system would have to bear in order to prevent further degradation of critical and non-substitutable natural capital and, then, to make further estimations of the expenditure necessary to move that "steady state" back to more sustainable levels of activity. Even under the most conservative of estimates, these expenditures are so large as to have massive impacts on western levels of physical consumption. Indeed, Hueting and Ekins (see endnote 11) have argued that we should talk in terms of more or less un-sustainable activities in order to avoid detailed concerns about what sustainability really looks like. We know we are moving towards greater and greater levels of un-sustainability. The first task is therefore to start to reduce that un-sustainability. This will be massively expensive and disruptive. Furthermore, this is without addressing thorny matters of eco-justice and more equitable distributions of consumption, wealth and environmental impact.

C. The sustainable business

If business is to be part of the process of re-directing economic activity away from increasing un-sustainability then it may be useful to assess whether business can be sustainable under present economic arrangements. And if so, how? It follows from the above analysis that, at a minimum, a sustainable business is one which leaves the environment no worse off at the end of each accounting period than it was at the beginning of that accounting period. For full sustainability, the sustainable business would also re-dress some of the excesses of current un-sustainability and consider the intra-generational inequalities. It is perfectly clear that few, if any, businesses, especially in the developed economies, come anywhere near to anything that looks remotely like sustainability.

Environmentally aware companies have stated within their publications opinions such as:

"... very few businesses can be said to be sustainable in their operations, since in one form or another their activities consume non-renewable resources, whether directly through travel or their use of raw materials

or indirectly through their use of energy."
(IBM UK, Progress Report, 1993, p.38)

"We challenge the notion that any business can ever be 'environmentally friendly'. This is just not possible. All businesses involve some environmental damage. The best we can do is clear up our own mess while searching hard for ways to reduce our impact on the environment."
(Body Shop, Green Book, 1992)

"Few, if any, businesses could claim to be anywhere near this position [of sustainable development] at present, but a number, including BT, have begun to consider what it might mean."
(British Telecom, Environmental Performance Report, 1993, p.35)

"I strongly believe that economic development and environmental protection should be pursued as mutual goals, but I have real doubts about the extent to which any company can manage its operations on a truly sustainable basis."
(Thorn EMI Environmental Report, 1994, p.4)

"At first sight the concept of sustainable development seems straightforward. But if its implications are examined, it becomes clear that a fundamental rethinking of attitudes towards production and consumption is required on the part of companies, society and the individual."
(Ciba Geigy Ltd, Corporate Environmental Report, 1992, p.14)

These statements contrast markedly with the casual use of the term "sustainable" which one will find in the environmental reports of a number of transnational corporations (TNCs) which assert to such matters as their "sustainable forestry policies", or "sustainable use of resources" or the "pursuit of sustainable resource use through constant improvement in efficiency" and so on. These are not statements about sustainability and mis-use through mis-understanding, an important and seriously challenging concept.

A particularly clear case for the un-sustainability of business is expressed by Dr. Chris Dutilh of Van den Bergh (Unilever) Netherlands. Dr. Dutilh illustrates the case by reference to two diagrams -- shown here as Figures I.1 and I.2.

Figure I.1 attempts to recognise the full environmental impact of different economic/ business activities. (This full impact could equally well cover eco-justice issues as well). Some proportion of the environmental impact is under the control of the company or industry concerned (for example, the use of energy) whilst some of the

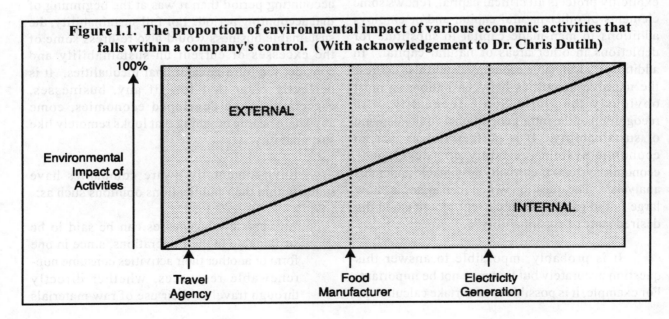

Figure I.1. The proportion of environmental impact of various economic activities that falls within a company's control. (With acknowledgement to Dr. Chris Dutilh)

impact is (currently at least) outside the business' control (for example, supplier, customer and disposal aspects). The proportion which is within or without the business' control obviously varies considerably between economic activities. Three possible examples of these are shown along the bottom of Figure I.1. For example, a travel agent will have very little direct impact on the environment although the activity of tourism itself has a massive impact through the activities of operators. On the other hand, food manufacturing has a major impact within the company's control but there is still a considerable degree of impact represented by the activities of (for example) agricultural suppliers and the actions and lifestyles of consumers. Finally, electricity generation incurs most of its environmental impact on the site of generation and a lesser impact is outside the company's control in the form of, for example, supply of the fuel[14]. Any company, by itself, can control only a small proportion of the total impact. This is shown in Figure I.2. The company can manage through EMAS-related control and innovation a substantial minority of the overall environmental impact of its activity. This proportion is greater, the greater the proportion of the total impact that lies within the company's control. The motivation for these EMAS-related controls and innovations will, on the whole, be economic -- of the "pollution prevention pays" variety. Further reductions in environmental impact are possible through supply-chain initiatives and

through educating the consumer of the product. However, together these two initiatives can still only reduce a relatively small proportion of the activities' overall environmental impact. The rest of the impact lies outside the control of the company. In particular, it lies in the hands of governments, individual employees, individual consumers, overall lifestyles and, at a societal level, general aspirations and expectations.

There is, probably, a further layer of influence that could be added to Figure I.2. This would relate to the company's own influence on consumption patterns and tastes through activities such as advertising, style-changes and lobbying. However, many companies argue that no single company can pull back from, for example, encouraging people to consume its products as this will probably leave the company at a major economic disadvantage *vis-a-vis* other companies in the industry. This issue, most companies would argue, is also a case for society to decide. Society must set the terms for business. Business will respond.

Taken from this perspective, then it is wholly unreasonable for society to expect business to become sustainable under its own initiatives. If economic systems are to move towards sustainability, then the "rules of the game" will have to change. How much influence companies have on that process is largely a question of one's perspective. It is perfectly clear that business is

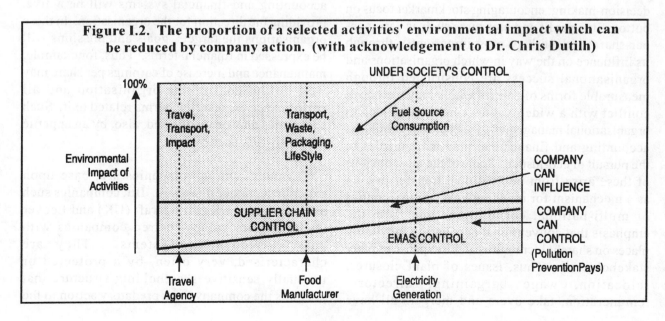

Figure I.2. The proportion of a selected activities' environmental impact which can be reduced by company action. (with acknowledgement to Dr. Chris Dutilh)

not, and cannot be, sustainable under present economic conditions. The issue to be investigated is not, therefore, with business sustainability, *per se*, but the steps that companies are taking to reduce their un-sustainability and the extent to which they are aware of the problems that the path towards sustainability is placing in their way.

It is in this context that accounting has several important roles to play.

D. Accounting for sustainability? Accounting towards sustainability?

Accounting in its broadest sense[15] can play, and does play, a number of roles in the company's pursuit of sustainability. Some of these roles are positive (in the sense of encouraging the corporation to move towards less-unsustainable actions) whilst others may be negative (in the sense of sending signals which encourage environmentally-malign behaviour and in offering resistance to new initiatives designed to encourage more socially and environmentally sensitive behaviour).

The negative roles that accounting can play have been widely described[16]. Broadly speaking conventional accounting and finance can hold back the pursuit of eco-efficiency through means such as: restrictive budgeting practices; short-term profit-centred performance measurement; overly restrictive capital budgeting and new product decision-making; encouraging stockmarket focus on bottom-line profit and measures such as earnings-per-share; and, perhaps most importantly, through its influence on the way in which organisations and organisational success are constructed through measurable forms of control which may well be in conflict with a wider, values-centred approach to organisational management, (see below). Similarly, accounting and finance can play malign roles in the pursuit of eco-justice. Some of the most obvious of these arise from international transfer pricing as a mechanism for reducing host country control of multi-national corporations and from the emphasis that conventional accounting inevitably places on shareholder returns at the expense of other stakeholders[17]. Thus, issues of plant closures, relocation, wage bargaining, directors' remuneration, take-overs and mergers all have implications for the distribution of wealth both within developed economies and between developed and under-developed economies.

On the other hand, accounting can play a very positive role in helping companies develop environmentally and socially sensitive policies. To date these positive roles have tended to be less developed than the negative roles, but increasingly companies (and accountants) are recognising how accounting and financial systems can be employed in innovative ways. These are broadly summarised in Table I.1.

However, there is one final point to stress before turning to look at the data collected. It is not necessarily the case that accounting and financial systems play an all-embracing role in an organisation. Indeed, it is quite possible that an organisation serious about sustainability would take every step to reduce the influence of the conventional accounting and financial systems. One way of recognising this is shown in Table I.2. Most traditional and large companies will probably be "performance-centred" with an active accounting system. In some, the accounting and financial systems will be more passive, responding and reactive rather than proactive. Whilst the financial systems will record "success/failure", the targets, objectives, goals, and missions of the company are likely to be expressed more in terms of growth, market share and quality statements. In others, the accounting and financial systems will be active, providing the direction for the organisation. In these organisations the targets, goals and missions will be expressed in financial terms. Thus, for example, maintenance and increase of earnings per share may well be the goal of the organisation and all performance expressed in terms related to it. Such companies may be recognised, also, by an appetite for profit-rich acquisitions.

Values-centred companies are, base upon experience, rather more rare. But companies such as Body Shop (UK), Traidcraft (UK) and Ecover (Belgium) are values-centred companies with passive accounting systems. They are characterised, very often, by a protected or unusually sensitive shareholding structure that prevents the company from predatory action in the

TABLE I.1. Roles of accounting in the pursuit of sustainability

STAGES OF ORGANIZATIONAL DEVELOPMENT	ECO-EFFICIENCY ISSUES	ECO-JUSTICE ISSUES
Improvement within current economic orthodoxy (Reducing un-sustainability):	* EMAS accounting - for wastes, efficiency, energy, pollution, etc. ("Pollution Prevention Pays"); * Reworking investment appraisal methods; * Contingent liabilities, asset revaluation and other financial reporting issues; * Tellus Institute methodology; * Basic environmental reporting.	* Employee and employment reporting, information for collective bargaining; * Value-added statements; * Bilan Social; * Community reporting; * Stakeholder analysis.
Recognition of the demands of sustainability	* Sustainable cost calculation and reporting; * Full cost accounting (EU 5th Action Plan); * Advanced environmental and sustainability reporting (including LCA and oköbilanz) - accountability and transparency.	* Full social reporting and social bookkeeping systems; * External social audits; * Transparency on transfer pricing and resource acquisition issues; * Accountability and transparency.

TABLE I.2. Role of accounting in the company's environmental response

TYPE OF ACCOUNTING SYSTEM	TYPE OF COMPANY	
	Performance-Centred (Accounting often dominant)	Values-Centred (Accounting less important)
Accounting System Passive	Driven by stock market, profit, financial performance appraisal	Other mechanisms used to identify financial efficiency - if at all
Accounting System Active	Methods of accounting explicitly linked to company's goals	Actively part of management process - BUT very constrained

stockmarkets. Whilst it is difficult to be certain, the actions of Ben and Jerry's (United States) -- a company widely recognised as values-centred -- appears to have an active accounting and financial system in that the development and expansion policy of the company appears to be tightly controlled in financial terms but, on the surface at least, the financial systems appear to be integrated with, and subservient to, the company values[18].

The outcome of these differences means that, in the values-centred company any development of the accounting and financial systems to encourage more environmentally and socially-sensitive behaviour may well be counter-productive. The values of the company will drive behaviour in these directions and the accounting systems will simply follow as required. On the other hand, in the performance-centred companies, it is highly unlikely that much progress can be made on either the eco-efficiency or eco-justice fronts without the developed support of refined and improved accounting and financial systems.

E. Investigating the issues

These were the background issues which informed and determined (a) the way in which the questionnaire was developed and subsequently used and (b) both the selection procedure for companies which form the case-studies and the way in which discussions were held in the companies. It is in the context set out above that the results are interpreted and, as shall become apparent, certain of the responses in the questionnaires might have led to alternative interpretations if the analysis started from a different point of view. As with all data, the researchers' orientation as described above should be noted when drawing conclusions. (More details on the method and mechanisms of the research are given in Appendix A, which is a separate document and is available only in the English language.)

II. QUESTIONNAIRE RESULTS

Although questionnaires were received from companies in different countries, from different industry sectors and of a variety of sizes[19] there was notable commonality of views about aspects of the nature of sustainability, what it meant for the business and the role that accounting can, and could, play in the pursuit of sustainability. This section reviews these results and follows the structure of the questionnaire itself (see Appendix B which is also a separate document and available only in the English language).

A. The nature of sustainability

Respondents were offered a range of statements concerning the nature and implementation of sustainability in business and were asked to indicate their agreement or disagreement with these[20]. There were 8 of these statements with which 75 per cent or more of the respondents could agree. These are shown, with the percentage agreeing, in Table I.3.

Respondents also seemed to be in broad agreement on the compatibility of sustainability with economic growth (67 per cent) and that businesses' pursuit of sustainable development involved: consideration for people of developing countries (74 per cent); commitment from the financial markets (70 per cent); and, concern for wealth distribution between nations (66 per cent).

On the other hand, there was also evidence that business (as a whole) does not hold a single, clear interpretation of sustainability. Two factors point to this. First, more than one quarter of the respondents were undecided about issues such as: the role of environmental management systems; the role of accounting systems; the magnitude of the costs involved; the compatibility of sustainability with economic growth; and, the extent to which sustainability was currently achieved in the respondent's business. This indecision is important because it highlights that a number of organisations lack direction in their search for a sustainable path.

TABLE I.3. Views on the nature of sustainability

Business' pursuit of sustainable development.....	Percentage agreeing with the statement
..requires a partnership approach from government, business and society	96
..means tackling both social and environmental problems	86
..requires balancing the needs of the economy with environmental protection	86
..is compatible with the profit ethic	82
..will involve all parts of society in discussion and implementation	82
..involves fundamental changes in attitudes and values	82
..is possible	78
..is likely to be complex and require extensive changes	75

Second, and far more persuasive, is the number of companies which believe: that sustainability does not involve the needs of future generations (59 per cent); that sustainability is synonymous with environmental management systems (45 per cent); and that their organisation already achieves sustainability (37 per cent). Such beliefs seem completely at variance with the most basic and commonly accepted parameters of sustainability.

The essence of the enquiry in this question concerned three basic issues or groups of issues: (a) did companies see sustainability as compatible with the status quo?; (b) did companies appreciate the eco-justice as well as the eco-efficiency issues?; and, (c) did companies appreciate the difficulty of the challenge set by the concept?

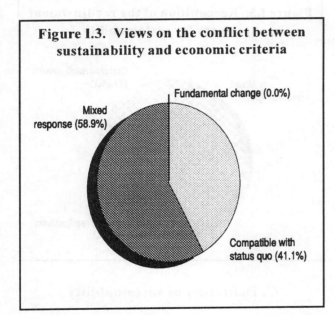

Figure I.3. Views on the conflict between sustainability and economic criteria

If sustainability is compatible with the status quo it will be "achievable", compatible with profitability and compatible with economic growth. Forty-one percent of the companies agreed that sustainability was all three. This is shown in Figure I.3 as seeing sustainability as "compatible with the status quo". On the other hand, no companies saw sustainability as it is usually typified in the literature as being exceptionally difficult, incompatible with both economic growth and the profit ethic and probably involving something other than a "balance" between economy and the

environment. Most companies (59 per cent) gave mixed responses to these questions.

Secondly, if companies recognised the importance of the eco-justice issues then, within question 1 they could be expected to agree broadly with questions relating to partnership issues, accountability and transparency, future generations, wealth distribution and third world issues. The proportion of companies subscribing to these views is shown in Figure I.4. The majority of companies display a consistently positive attitude to those aspects we have labelled "strong" eco-justice. That is, they appear to be highly aware of and supportive of the social requirements of sustainability. On the other hand, very few companies expressed views entirely at variance with a recognition of the eco-justice point of view. (This 1.4 per cent is shown as "weak eco-justice" in Figure I.4). Nearly half of the respondents (45 per cent) gave a mixed response on this issue.

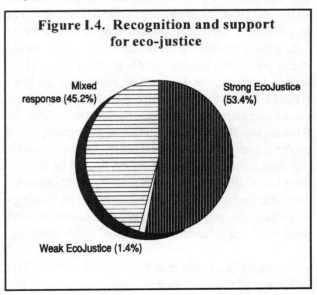

Figure I.4. Recognition and support for eco-justice

Finally, a company which had a good grasp of sustainability would appreciate that the path towards sustainability would involve changes in attitudes and values, would be complex and probably costly. The 37 per cent of companies subscribing to this view are shown in Figure I.5 as recognising that a "complex readjustment" is needed. No companies expressed views that indicated the transition would be entirely straightforward ("simple readjustment"). The majority of companies had a mixed response to the issues.

It is difficult to draw clear conclusions from this question. However, the data suggests that although there are signs of a lack of sophistication in companies' understanding of the implications of sustainability, respondents do appear to recognise that eco-justice issues are relevant and also to recognise that the path towards sustainability is likely to be complex.

B. Influences on understanding sustainability

The sources of respondents' understanding of the concept are important because sustainability is both a difficult and emerging concept. Different sources give different insights into the issues and emphasise different aspects of how to move towards sustainability. Furthermore, knowledge about the sources that companies consult may have implications for those bodies who are seeking to help business develop its posture of response to the issue of sustainability. Over 70 per cent of respondents were influenced by, *inter alia*, the ICC, Agenda 21 and the Rio Summit, their own company, books and economic journal articles, the media, the Brundtland Report, their national government and professional or trade associations. Of these, easily the most influential were the first four. The BCSD, the EU's Fifth Action Programme[21], environmental pressure groups, the World Conservation Strategy and other non-governmental organizations (NGOs) were the least influential sources of understanding.

Some inference may be drawn from this. The influence of the respondent's own company is not unexpected but, given that the questionnaire was completed by the member of that company who understood the issues best, the influence is confusing. That is, (a) there is no *a priori* reason why a company should have a sophisticated understanding of the issue; and (b) if anybody within the company did understand the issue it would be the person completing the questionnaire. It is likely therefore, that the influence of the company was concerned with whether or not initiatives towards sustainability could be taken -- rather than an influence on the understanding itself.

Agenda 21 would not have provided much direct guidance to a company seeking sustainability but, clearly, raised the issue to a significant degree in the global context. Its influence seems, again, to be one of awareness and changes of the political agenda for the environmental management team. So the principal influences on understanding are the respondents own self-education through books and journals and the ICC. It was expected that the influence of the ICC (and, less correctly, the BCSD) would be high. The fact that the ICC does not actually address sustainability goes some way towards explaining what look like naïve understandings of the concept amongst many of the respondent businesses.

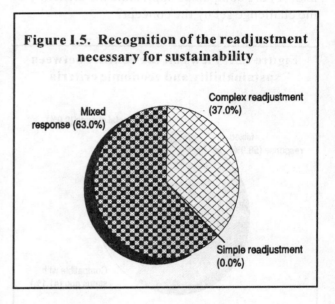

Figure I.5. Recognition of the readjustment necessary for sustainability

Mixed response (63.0%)

Complex readjustment (37.0%)

Simple readjustment (0.0%)

C. Definition of sustainability

Commentators have argued[22] that the very general nature of the definition of sustainability has been instrumental in permitting widespread agreement about its desirability. A more specific definition might raise conflicts over, *inter alia*, issues of redistribution of wealth and challenges to existing western patterns of development and consumption. And yet previous work[23] has identified that companies need specific guidance on the implications of the term if they are to address the issue directly. Over 60 per cent of respondents said they favoured a more specific definition of the term. Later questions also addressed this issue and a substantial minority of respondents (36 per cent) suggested that the lack of precise definition was a real constraint on their moves towards achievement of sustainability. It is not clear to what extent companies who recognise the need for more precision in the application of the term appreciate the implications this may have for current economic orthodoxy. For example, it is apparent that some of the best-informed respondents did not want a specific definition of the term because they did appreciate its potential implications for business practices whereas some of the least-informed wanted a less-general definition simply because they did not appreciate the extent to which the pursuit of sustainability involved far more than sophisticated EMAS. If we believe that development of EMAS is a useful step towards sustainability then a more specific definition may be premature. On the other hand, the implications of the term require very careful debate and examination and this may be a matter that policy-making bodies should be addressing at as early a stage as possible. In the absence of a specific definition, sustainability may be becoming a "motherhood" term whose parameters are left in the hands of industry groupings such as a nation's confederations of business, the ICC and the BCSD. This is certainly practicable in the short term. It may not be effectual in the long term.

D. Formal recognition of sustainability in the company

Eighty-two percent of respondents said that their company formally recognised sustainability. Ironically, the majority of these "formal recognitions" did not explicitly cite sustainability or definitions of sustainability. There was, on the whole, recognition of the environmental agenda within policy statements as well as recognition of the need for development of EMAS and related initiatives. In less than 20 per cent of cases could some explicit recognition of the term be offered as evidence of the formal recognition by the company. This, again, suggests that a high proportion of companies have yet to appreciate that general environmental policy statements coupled with EMAS does not guarantee sustainability. This was recognised by the majority of the companies which replied that their organisation did not formally consider sustainability. Eleven percent of

respondents described their (significant) steps towards environmental management but explicitly recognised that these are not necessarily sufficient for a formal recognition of sustainability. One of the very few clear country exceptions was apparent here. Half of the Dutch companies clearly understood sustainability but clearly recognised that their advancement of EMAS did not guarantee its achievement.

E. Encouraging moves towards sustainability

Easily the biggest influences on respondents in "their moving towards sustainability" were the laws (actual and impending) and market opportunities. Over 86 per cent and 90 per cent of companies, respectively, cited these influences. This more than anything, illustrates the potential confusion over the meanings of the term "sustainability". Whilst the term has now crept into some areas of some national laws, there is no evidence that current or immediately impending laws actually addresses the structural changes that sustainability will require. Furthermore, there are remarkably few market opportunities which can be demonstrated to be commercially attractive and inherently sustainable. What we know from elsewhere[24] is that these influences are significant in encouraging companies on the path towards better environmental stewardship and management. Similarly, the third and fourth sources of encouragement -- the company's national government (79 per cent) and employees (74 per cent) -- have been widely recognised as important influences on a company's environmental response. Other sources of encouragement towards sustainability included the European Union (70 per cent), industry associations (64 per cent)[25], customers (64 per cent), NGOs (58 per cent) and pressure groups (51 per cent).

The increasing doubt which the questionnaires engendered as to whether or not most companies appreciated the full implications of sustainability were further reinforced by this question. That is, very few companies identified any of the sources as discouraging moves towards sustainability. This is clearly incorrect.

This was also the first stage at which accounting and financial systems entered the enquiry. Whilst a substantial proportion of respondents considered that capital markets (49 per cent), the financial accounting system (67 per cent) and financial institutions (73 per cent) played no part in encouraging or discouraging moves towards sustainability, these three were the principal sources identified by companies as discouraging sustainability. (See also part G below). Eight percent believed that capital markets and financial institutions discouraged them, whilst 14 per cent identified the financial accounting system as a major discouragement of any moves towards sustainability.

F. Activities which encourage moves towards sustainability

Respondents were offered a range of activities and asked to identify which might move their organisation towards sustainability. Table I.4

summarises some of the results from this question[26]. In broad terms, companies, as expected, are generally well-informed about eco-efficiency and environmental management issues. The level of activity in these areas, whilst no cause for complacency, is also broadly in line with other surveys.

Companies are less enthusiastic about accountability and transparency issues. But even here, there is something approaching agreement on the importance of openness. Notably, redesign of the accounting system was the area, companies considered, that would contribute least to the pursuit of sustainability. More significantly still, it was the two specifically accounting-related areas of investment decisions and developments of accounting systems on which companies were least sure on how to proceed. This seems to point to a need for wider dissemination of proposals and further practical research work into the issues.

TABLE I.4. Importance of selected activities on moves towards sustainability

	Percentage of Companies Responding			
	Will contribute towards sustainability	Will not contribute towards sustainability	Company is active in the area	Company not sure how to proceed
Reduction in energy usage	96	1	74	0
Reduction in the consumption of resources	96	1	67	0
Reduction in the volume and toxicity of wastes	85	3	67	1
Targets for moving towards sustainability	84	3	48	8
Recycling	82	4	68	3
Commitment to an environmental management and audit system	81	7	66	4
Use of social and environmental criteria in capital budgeting and investment decisions	74	5	40	21
Reporting on the achievement of sustainable development targets	74	4	48	8
Product design to increase the life of the product and improve reparability	74	11	41	10
Providing information to shareholders and financial community on your social and environmental impact	67	4	55	10
Policy statements regarding the pursuit of sustainable development	64	10	56	3
Being open to scrutiny by groups and individuals outside of the organisation	56	15	47	4
Redesigning the accounting system -e.g. for full cost accounting or for environmental provisions	51	14	19	23

G. Constraints on the pursuit of sustainability

Whilst 16 per cent of respondents replied that sustainability was unimportant to the organisation and 23 per cent suggested that there was no demand for the company to pursue the concept, most companies which identified constraints saw the principal constraints they face as comprising: a lack of guidance (47 per cent), clear definition (36 per cent) and support (53 per cent); a problem of priorities (51 per cent); and accounting-related problems (47 per cent). Probably most importantly, however, company views on the issue of constraints were widely dispersed. There is no unanimity on this issue. Virtually half of the respondents saw few, if any, constraints on their capacity to pursue sustainability. Those who did see constraints wanted further government support and guidance and/or saw that current measures of success (e.g. growth and profit) and the present system of accounting as their principal impediment[27]. At this stage it is not possible to determine possible explanations of why these differences in views arise. At least part of it, we infer, is related to the understanding of sustainability but there are likely to be other issues that will require further analysis to isolate. There would certainly appear to be policy-implications in even these initial findings.

H. Accounting and financial systems issues

As might have been expected, accounting and finance have not appeared as central issues in companies' responses about sustainability. Most companies have not seen profit or the capital markets as either major problems or potential sources of conflict. As a result, it is no surprise to discover that accounting is one of the least developed areas of companies' response to the sustainability agenda. And yet, accounting is clearly an area on which a substantial minority of companies need further guidance and where significant future developments and problems may emerge. It was central to this project to try and investigate these matters further.

Table I.5 summarises the dominance of the accounting system in company management. The diversity of views amongst company respondents is again apparent in this figure. While a majority of companies are accounting-dominated in traditional areas of decision-making, practice varies considerably in other areas -- especially those related to social and environmental matters. This diversity of experience may be part of an explanation as to why respondents were less unanimous on the role of accounting in the pursuit

TABLE I.5: Accounting's dominance in company decision-making

	Percentage of companies who indicated that accounting and finance systems and criteria			
	..dominate organizational decisions	..are just one set of criteria	..are dominated by other criteria	No response
Medium and short term planning	78	15	5	2
Capital expenditure and investment	75	18	5	2
Divisional performance evaluation	71	22	5	2
Corporate strategy	59	34	5	2
Managerial performance evaluation	52	36	10	2
Relationships with the financial community	51	32	16	1
Purchasing decisions	47	45	7	1
Make or buy decisions	41	48	7	4
Remuneration of senior executives	33	40	25	2
Environmental projects	30	42	26	2
Environmental standards and targets	27	37	34	2
Research and development initiatives	26	41	32	1
Staff training	22	49	27	2
Community involvement	16	36	45	3

of sustainability -- in less accounting-dominated, more values-centred organisations, management may well look to other measures to appraise and assess options without keeping one eye firmly on the "bottom line". This, again, deserves further investigation.

Which brings us to two of the central questions which this research sought to answer -- "what innovative accounting methods are being used by companies?" and "why?". The results are not especially encouraging at this stage of development. Only a minority of respondents are familiar with a number of the accounting options devised to help a company assess its path towards sustainability. A further minority is actually experimenting with these options. If accounting-related techniques do have a role to play in helping companies redirect their activities -- and it seems increasingly certain that they do -- more effort needs to be exerted in disseminating these proposals and encouraging companies to experiment with them. Other evidence suggests that both LCA and mass-balance/oköbilanz techniques have real practical benefit for the companies themselves as well as highlighting the real issues facing a company considering addressing sustainability seriously. Other methods are less widely applied in practice as yet but deserve practical experimentation -- if only to either reject them or develop ways around their practical problems. There is clearly a need for more education of companies in this area.

Finally, the questionnaire offered companies a range of accounting-related activities which are known to have been successfully applied in organisations and which can contribute to both environmental sensitivity and economic success. In the light of this, the fact that only 75 per cent of respondents identified either accounting for energy costs and/or accounting for waste costs, and only 32 per cent are using accounting to help allocate environmental overheads, is disconcerting. We are looking here at some of the world's leading companies and they are not yet employing their accounting systems in those areas where the economic and environmental savings are the most pronounced and the simplest to identify. Identification of various costs was the most popular use of the accounting systems with, in addition to wastes and energy, EMAS costs (70 per cent), investment procedures (68 per cent), compliance costs (64 per cent) and packaging retrieval and disposal costs (56 per cent) the most popular. Other significant areas of accounting-related activity were environmental reporting (67 per cent) and identification of market opportunities (59 per cent).

The most striking results to emerge from this area, however, related to what companies did not see as important and/or on which companies were unsure as to how to proceed. It is now widely accepted[28] that any significant movement towards environmental sensitivity -- let alone towards

TABLE I.6. Companies employing or familiar with 'new' accounting methods

	Percentage of Respondents who..		
	Are using this is approach	Are familiar (but not using) this approach	Are unfamiliar with this approach
Life Cycle Assessment with financial numbers assigned to physical flows	16	15	48
Extended cost benefit analysis	14	10	48
Full cost accounting (e.g. the Tellus Institute)	11	14	51
Eco-balance sheets/Oköbilanz (e.g. The Danish Steel Works) with financial numbers assigned	10	16	58
Net environmental value added calculations (such as that done by BSO/Origin)	8	4	74
Full cost pricing (e.g. EU Fifth Action Plan)	7	14	52
Social Auditing (e.g. Traidcraft or Sbn Bank)	7	0	77
Sustainable cost calculation (e.g. CSEAR)	5	3	78

sustainability -- requires a link to be drawn between the economic (financial accounting) performance and environmental performance of a company. The first step in this direction involves conventional financial accounting adjustments. For very few companies indeed are these matters unlikely to be important. However, a significant minority of respondents saw such matters as "not important/ not sure how to". These included: disclosure in financial statements (25 per cent unimportant/14 per cent not sure how to); environmental contingent liabilities (18/19 per cent); environmental provisions (27/14 per cent); revaluation of land and inventory (19/12 per cent); relationships with shareholders and the financial community (21/4 per cent); meetings with analysts and ethical investment trusts (36/10 per cent); and liaison with the financial auditors (37/14 per cent).

Whilst some of these views on the lack of importance of financial accounting issues (as well as lack of clarity on procedure) could be traced back to (a) countries with either a less active stock market or an accounting profession which has taken less of a lead on accounting for the environment and (b) industries which perhaps might be expected to be less exercised by such matters, this was by no means true in all cases. Indeed, a number of the respondents confirmed that many companies are either unaware of the financial issues that may attach to environmental matters or else they are simply waiting until their national government takes a firm lead and provides guidance on the treatment of these environmentally-related financial accounting questions.

III. CONCLUSIONS

This report is essentially an exploratory piece of research undertaken on a new and emerging area where little prior work exists to provide guidance. Based, as it is, on a range of postal questionnaires and cases, it cannot be regarded as representative of any country or industry. Nevertheless, thanks to the selection procedure used to identify companies, it is hoped that there is, in the results, a useful picture of how many of the world's more environmentally advanced companies are reacting to the emerging issue of sustainability. The majority of the questionnaires were completed by senior

environmental executives and this should provide a flavour of how the top levels within leading businesses are thinking about the issue. However, it may be that such respondents were unable to answer as fully about some of the accounting issues as, say, a financial director might have done. The cases do not suggest that this is so. The cases (and some previous work[29]) suggest that accounting functions are more usually passive and often await direction from other areas of the business. This seems in accord with the results produced here.

With this in mind, the following tentative conclusions can be drawn:

(a) Most companies -- even leading companies -- have yet to examine fully the nature of sustainability and its implications for current business practices. The vast majority of companies are treating sustainability as an (implicit) extension of environmental management. It is highly unlikely that environmental management can provide sufficient impetus towards sustainability. It is almost certain that the 37 per cent of the respondent companies which believe that they are at or close to a sustainable state are quite simply mis-informed.

(b) While companies do not appear to have fully embraced all the eco-efficiency implications of sustainability, they do seem sensitive to and supportive of the eco-justice elements of the concept. However, the question of who has responsibility for addressing eco-justice issues emerged from the cases as a problematic area. This recognition is not uniform. A surprising 59 per cent of companies appear to believe that sustainability does not involve the needs of future generations. This is also a mis-informed view.

(c) The dominant conclusion which was drawn from this study is that further business-related research needs to be undertaken and much clearer guidance given to businesses world-wide on what sustainability may mean for business.

(d) Environmental managers are constrained in the actions they can take to move their companies along the path to sustainability. Traditional accounting comprises an important part of this constraint. The resultant conflict is more likely to occur in accounting/performance-centred organizations than in values-centred ones. It may also be more likely to occur in very large, centrally controlled organizations than in smaller or decentralized companies. This is, however, a very tentative conclusion.

(e) The use of LCA and/or the ökobilanz approach to understanding a company's interaction with society and the environment seems to offer considerable promise. Not only will this aid companies in developing their environmental management systems but it will also help the company to identify and confront the issues of sustainability.

(f) The central messages about the assistance that accounting systems can provide to companies addressing environmental issues has not yet been fully received. Much more can be done in (i) areas where environmental and economic savings can be made; and (ii) recognizing those areas of financial reporting where environmental costs must be identified and reported. Accounting needs to be fully integrated into a company's EMAS strategy.

(g) Other, more innovative approaches to accounting and sustainability need to be further researched, applied and disseminated. It is increasingly clear that there is value -- both economic and environmental -- in these approaches and while such approaches may not always produce the answers that a business (understandably) wants, sustainability -- and forward-looking business management -- demand a re-think of current business orthodoxy. Very few companies are active in the use of these techniques and not many more are actually familiar with them. For example, the EU must do a great deal more to explore and disseminate the implications of any approach to full cost pricing. It does not appear that this message has been understood as yet.

(h) It needs to be recognized that companies cannot be sustainable in the present economic climate. There is widespread agreement that the costs of sustainability-related adjustment will be considerable and to ask any company to act entirely as a philanthropist and/or avoid pursuit of

traditional business opportunities is quite unrealistic. Businesses are successful at growth, efficiency, innovation and exploitation of markets. To expect such successful organisms to change their approaches -- without significant changes in the business environment -- is too optimistic. It is highly unlikely that any companies can be sustainable under present economic strictures. If sustainability depends on changes taking place within the current business orthodoxy, then it is not a realistic possibility.

(i) The pursuit of sustainability will involve all parts of society. Voters, consumers, employees, management, politicians etc., will have to address the issues as they relate to their own sphere of influence. Likewise, business and governments will have to confront challenging and difficult choices. In particular, national and pan-national governmental organizations must address sustainability firmly and produce an overview and guidance which only they are equipped to provide. Only then, within this wider context, may any realistic expectation of sustainable businesses be articulated and realized.

Notes

1. United Nations World Commission on Environment and Development. Our Common Future (The Brundtland Report). Oxford: Oxford University Press, 1987.

2. The two terms "sustainability" and "sustainable development" are used synonymously in this report although, for some, they have different connotations.

3. "Towards sustainability". Com(92) 23 final - Vol.I-III, Brussels, 27 March 1992.

4. The International Chamber of Commerce. Business Charter for Sustainable Development. Paris: ICC, April 1991 and first released at WICEM II.

5. EMAS, as the shortened version of the European Union's Environmental Management and Audit Scheme, will be used generically throughout this report to cover environmental management and environmental management systems -- including environmental audits.

6. An important and frequent fallacy is the assumption that efficiency is likely to lead to sustainability. As a company increases its efficiency with respect to resource use, it decreases its *unit* impact on the environment. Companies are a very long way from reducing this impact to either zero or, at least, to the carrying capacity of the eco-system. If the company is simultaneously expanding and seeking growth -- as most successful companies tend to do -- then, despite the efficiency, the company's impact on the environment is very probably continuing *to increase*, not to decrease. Hence, decreasing per unit impact does not assure an overall decrease in impact.

7. See for example Chapter 14 of Gray, R.H., K.J. Bebbington, and D. Walters. Accounting for the Environment: The greening of accountancy Part II. London: Paul Chapman, 1993.

8. op. cit.

9. Pezzey, J. Definitions of Sustainability, (No.9). United Kingdom: CEED, 1989.

10. The issues of eco-justice and eco-efficiency are explored further in Gladwin, Thomas N. "Envisioning the sustainable corporation" in (ed.) Emily T. Smith. Managing for Environmental Excellence. Washington, DC: Island Press, 1993.

11. This brief summary draws especially from the work of Pearce and Turner see, for example, Pearce D. (ed.). Blueprint 2: Greening the World Economy. London: Earthscan, 1991; Pearce, D. "Towards a sustainable economy: Environment and economics". The Royal Bank of Scotland Review, No.172, December 1991, (pp.3-15); Pearce D., A. Markandya and E.B. Barbier. Blueprint for a Green Economy. London: Earthscan, 1989); plus that of Paul Ekins and Roefie Hueting (see, for example, Hueting, R. New Scarcity and Economic Growth: More Welfare through Less Production? Amsterdam: North Holland, 1980; Hueting, R., P. Bosch and B. de Boer. Methodology for the calculation of sustainable national income. Voorburg: Netherlands Central Bureau of Statistics, 1991; Ekins, P. A New World Order: Grassroots movements for global change. London: Routledge, 1992; Ekins, P. Wealth beyond measure: An atlas of new economics. London: Gaia, 1992).

12. This is explicitly point of view of man as the centre of existence. For a further discussion of this issue in an accounting context, see Gray, R.H. "Accounting and environmentalism: an exploration of the challenge of gently accounting for accountability, transparency and sustainability". Accounting Organisations and Society 17(5), July 1992 (pp.399-426) and Maunders, K.T. and R. Burritt. "Accounting and Ecological Crisis" Accounting, Auditing and Accountability Journal. 4(3), 1991 (pp.9-26).

13. It is well established that the rate of environmental degradation is accelerating. Recent attempts to increase global environmental sensitivity have only, so far, managed to slow the rate of acceleration. Most measures of environmental degradation reflect this. Most measures of economic success also demonstrate a rising trend. Two series which are clearly moving in the same direction point to a high level of correlation and therefore present a *prima facie* case for a relationship between the two.

14. The fuel source would move the line to the left or the right depending on whether it was coal-, natural gas- or nuclear-fuelled electricity generation.

15. Whilst conventional accounting comprises the financial record-keeping and information systems for the management of an organisation together with the annual reporting of financial information for

shareholders and other stockmarket participants, accounting need not be so restricted. In

principal all flows of information -- whether financial or non-financial -- can be seen as a form of accounting. It is in this sense that we use the term here. See, for example, Gray, R.H. "Corporate Reporting for Sustainable Development: Accounting for sustainability in 2000AD". Environmental Values 3(1) Spring 1994, pp.17-45; Gray, R.H. and K.J. Bebbington. "Accounting, Environment and Sustainability". Business Strategy and the Environment. Summer 1993, pp.1-11.

16. See, for example, Bebbington K.J., R.H. Gray, I.Thomson & D.Walters "Accountants Attitudes and Environmentally Sensitive Accounting" Accounting and Business Research No.94 Spring 1994 (pp.51-75); Gray R.H., K.J.Bebbington, D.Walters & I.Thomson "The Greening of Enterprise: An exploration of the (non) role of environmental accounting and environmental accountants in organisational change" Critical Perspectives on Accounting (forthcoming);Gray R.H., D.L.Owen & K.T.Maunders Corporate Social Reporting: Accounting and accountability (Hemel Hempstead: Prentice Hall) 1987; Gray R.H., D.L.Owen & K.T.Maunders "Corporate social reporting: emerging trends in accountability and the social contract" Accounting, Auditing and Accountability Journal 1(1) 1988 (pp.6-20); Gray R.H., D.L.Owen & K.T.Maunders "Accountability, Corporate Social Reporting and the External Social Audits" Advances in Public Interest Accounting, Vol.4 1991 (pp.1-21).

17. There is some evidence that conventional shareholders are beginning to demand wider social and environmental information both through the ethical investment funds (see, for example, Harte, G., L. Lewis and D.L. Owen. "Ethical investment and the corporate reporting function". Critical Perspectives on Accounting 2(3) 1991, pp.227-254) and in more direct ways as environmental performance begins to excite shareholders (see, for example, Tilt, C.A. "The influence of external pressure groups on corporate social disclosure: Some empirical evidence". Accounting, Auditing and Accountability Journal, 7(4) 1994, pp.24-46 and Epstein, M.J. and M. Freedman. "Social disclosure and the individual investor". Accounting, Auditing and Accountability Journal, 7(4) 1994,

pp.94-109).

18. For another insight into this see, for example, Hawken, P. The ecology of commerce: a declaration of sustainability. New York: Harper Business, 1993.

19. More detail on this is given in Appendix A.

20. These statements were drawn from a wide selection of literature and were expressed in terms which varied the implied message of support or lack of support for various interpretations of the concept.

21. Italy was the only country in which the EU Fifth Action Programme appeared to be notably influential.

22. See, for example, Pezzey op. cit..

23. See, for example, Bebbington, K.J., R.H. Gray, I. Thomson and D. Walters. "Accountants Attitudes and Environmentally Sensitive Accounting". Accounting and Business Research, No.94, Spring 1994, pp.51-75; Gray, R.H., K.J. Bebbington, D. Walters and I.Thomson. "The Greening of Enterprise: An exploration of the (non) role of environmental accounting and environmental accountants in organisational change". Critical Perspectives on Accounting (forthcoming). detail).

24. Elkington, J. (with Tom Burke). The Green Capitalists: industry's search for environmental excellence. London: Victor Gollancz, 1987; Elkington J., P. Knight and J. Hailes. The Green Business Guide. London: Victor Gollancz, 1991; Davis, J. Greening Business: Managing for sustainable development. Oxford: Basil Blackwell, 1991.

25. National industry associations were cited by 21% of respondents and industry-specific associations were cited by 29% of respondents. Of the different industry groupings, the chemical industry associations exhibited the greatest encouragement. Japanese and South African companies were more likely to be influenced by their national industry associations.

26. The percentages for responses to will/will not contribute towards sustainability do not add to 100%. The differences are companies which were unsure.

27. Additional constraints identified by companies included the absence of a trained and educated workforce, existing legislation, difficulty of changing organisations, and, perhaps most importantly, a lack of will and direction from

society and government. Several mentions were made that society does not really want sustainability.

28. See, for example, Gray, R.H., K.J. Bebbington, D. Walters. Accounting for the Environment: The greening of accountancy Part II. London: Paul Chapman, 1993; Owen, D.L. Green Reporting: The challenge of the nineties. London: Chapman & Hall, 1992; Owen, D.L., R.H. Gray and R. Adams. "A green and fair view". Certified Accountant. April 1992, pp.12-15.

29. See, for example, Bebbington, K.J., R.H. Gray, I. Thomson and D. Walters. "Accountants Attitudes and Environmentally Sensitive Accounting". Accounting and Business Research, No.94, Spring 1994, pp.51-75; Gray, R.H., K.J. Bebbington, D. Walters. Accounting for the Environment: The greening of accountancy Part II. London: Paul Chapman, 1993; Gray, R.H., K.J. Bebbington, D. Walters and I. Thomson. "The Greening of Enterprise: An exploration of the (non) role of environmental accounting and environmental accountants in organisational change". Critical Perspectives on Accounting (forthcoming).

30. These activities were previously conducted by the former Transnational Corporations and Management Division of the Department of Development Support and Management Services.

31. This work built upon earlier work reported in the references to the work of Bebbington and Gray op cit.

32. This is probably the first recorded example in the accounting literature of such an extensive and innovative research project with such a wide range of international collaboration.

33. Returned questionnaires are still arriving at the time of writing but have not been integrated with the results reported here because of the time constraint. Copies of the report with the wider range of responses will be available from CSEAR in due course.

34. Further analysis of the results from the questionnaires will also be available from CSEAR in due course.

35. This case was put together from two site-visits and interviews, the first with the environmental manager and the second with the environmental manager and the chief accountant. The site visits also involved a tour of the factory and principal

UK site. Additional material has been gleaned from the company's published material including its environmental reports.

36. Case-2 was constructed from a site-visit of a CSEAR researcher accompanied by a local associate of CSEAR. The interview of nearly four hours with the head of environment for the world-wide foods division was conducted in English with occasional clarification in Dutch. A factory tour was also arranged. Additional information was gathered by the researcher through local knowledge and other contacts with the company.

37. This case was put together from an on-site visit by a CSEAR researcher, the local associate of CSEAR and an individual familiar with the company. Four individuals within the organisation were interviewed over a period of four hours with interviews being conducted in English and in Italian. Interviewees included the individual with responsibility for environmental reporting, the environmental manager, the chief accountant and the chief executive officer. Additional information was gleaned from the Annual and Environmental Report of the company.

38. This case was based upon a telephone interview with the director of environmental programmes. Unfortunately problems co-ordinating diaries and severe time constraints made a physical visit to this company impossible. Further information has also been obtained from (a) examination of the company's published material which is extensive and (b) discussion with a Fellow of CSEAR who has been involved with the company.

39. Case-5 involved a site visit by a CSEAR researcher to the company and interviews with the environmental manager and finance director over a three hour period. In addition a factory tour was arranged.

40. This case was constructed principally from a site visit and two interviews held, respectively, with the Director with Responsibility for the Environment and the Head of the Environmental Section of the company. The interviews were conducted in both English and Spanish with the local CSEAR associate present to interpret and explain issues as they arose. The visit lasted over three hours and was supplemented by information and explanations provided by the local associate who has a working relationship with the company.

41. Energy-generation and supply will frequently be an indigenous activity.

42. Its responses to the questionnaire on the sustainability questions were closer to the norm than other interviewed companies.

ANNEXES

Annex I

BRIEF OVERVIEW OF THE RESEARCH PROCESS

A. Background

The research reported here is a part of ISAR's continuing efforts to examine the development of environmental accounting and reporting by TNCs[30]. This report explores the role of accounting in the pursuit of sustainable development. This report represents the findings of research undertaken by The Centre for Social and Environmental Accounting Research (CSEAR) based at Dundee University. The university was also conducting research for the Chartered Association of Certified Accountants (ACCA) into accounting for sustainability. The CSEAR/ACCA project had involved both an extensive examination of published work on the nature of sustainability and its implications for business as well as a series of in-depth interviews with leading UK companies into their understanding of the nature of sustainability and what implications - if any - this might have for accounting practice[31]. Whilst it is clear from this work that it was essential to "invent" the *sustainable business* and to discover ways to account for such an entity, it was also apparent that companies had made little or no progress towards sustainability and, indeed, most companies were nervous about what signals an "account of their sustainability" might send to stakeholders at the present time.

This report investigates this issue in much greater depth and in an international context. More especially, the project was driven by a need to offer a greater voice to non-Anglo-Saxon commentators and corporations with particular reference to the European context of business, accounting and sustainability.

B. Intended structure of the project

The initial intention for the project was to focus on 5 non-Anglophone countries in Europe who could offer a contrast to the UK-based research already being undertaken on the CSEAR/ACCA project. A detailed questionnaire was devised and targeted at a selected few companies in each of the UK and the 5 other countries. This questionnaire would provide both an important element of this report and provide a basis for selecting an especially interesting company in each of the 6 countries for visits and case studies. The write-up of the cases would provide the second substantive element of this report.

C. The actual structure of the project

This report was undertaken with the assistance of an extensive network of International Associates, many of whom are the leading environmental accounting researchers in their own countries. The role of the International Associates was to identify the leading companies (with respect to sustainability) which were approachable on a project such as this in their own country. They would then negotiate with the company to establish the most suitable senior executive who would complete the questionnaire, determine whether or not the questionnaire would require translation and local-language testing. The questionnaire was circulated and returned. The International Associates would then negotiate interviews with senior executives in the most promising companies identified from the questionnaires. Visits and case studies would then be conducted by the CSEAR researchers accompanied by the local International Associate. This was a complicated and difficult project[32].

The questionnaire (see Appendix 3) was derived from the extensive research work and then field-tested through interviews with UK TNCs. Ten copies of the questionnaire were then despatched to Associates in each of 18 countries. Further questionnaires were despatched from CSEAR to 10 companies in each of the UK, Denmark and Switzerland (where CSEAR has no Associates at present). The wide coverage of the project was to (a) gain as diverse a range of opinion as possible, and (b) to allow for local difficulties - for example, refusal to complete questionnaires, difficulties in gaining actual access, problems with translation. This proved wise as: several countries were unable to partake in the research (for a variety of local reasons); the attitudes of senior executives to the project (and the completion of the questionnaire) varied enormously; and translation, despatch and completion of the questionnaire took significantly different lengths of time in different countries[33].

The questionnaires were returned together with background data on the companies provided by the Associates. These were collated and analysed[34]. The descriptive statistics from the survey are reported here. The national distribution of respondents to the questionnaire is shown in Figure I.A.I.1. An uneven national response is to be expected. Countries clearly vary in their willingness to respond to postal questionnaires and a number of countries were unable to gain any response at all despite existing contacts and repeated calls. However, returns are still being received and later interpretations of the data will not be so nationally uneven.

Figures I.A.I.2 and I.A.I.3 provide more background on the respondents. The companies represent a wide distribution of industry sectors and whilst none of the companies are small, there is a wide range of international size with only a very few companies (notably indigenous energy companies) operating only in their host country.

In only two countries were questionnaires returned from companies willing to take part in the second phase of the research (the visits) sufficiently early to permit the arrangement of interviews. So interviews were established by the Associates or directly by CSEAR. These were held during October and November 1994. These were of varying length but were all exceptionally useful.

D. Conclusions

An international project of such ambitious coverage - especially of a new and under-developed notion like sustainability - raised many unforeseen difficulties. Most of these have been overcome and the results can be treated with some confidence as offering a reasonable reflection of many of the world's largest and most environmentally-aware companies. Generally, views of the large companies shows a remarkable degree of agreement but a relatively low appreciation of the implications of sustainability and how to account for it.

Some suggestion of country-differences within Europe may be beginning to emerge but this deserves greater examination through more interviews. In addition, European views might most interestingly have been contrasted with those from other parts of the world. Future developments - and especially the EU Fifth Action Programme - also suggest that further work along these lines will prove to be very productive.

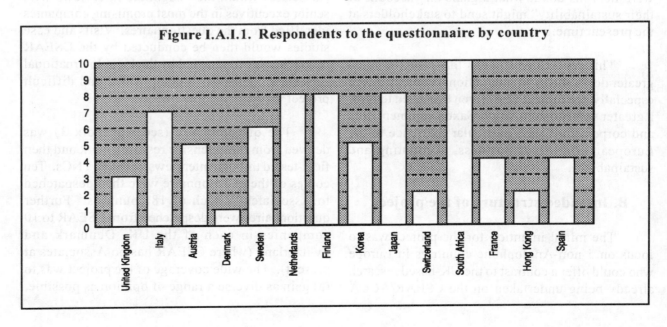

Figure I.A.I.1. Respondents to the questionnaire by country

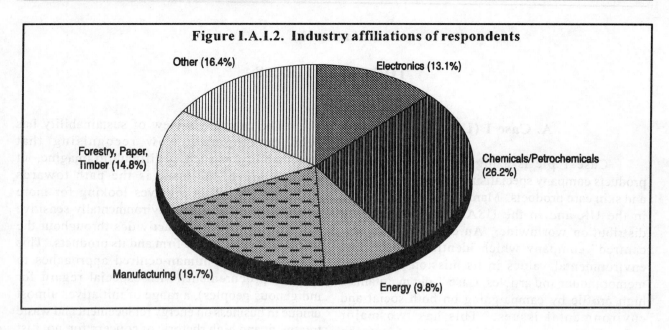

Figure I.A.I.2. Industry affiliations of respondents

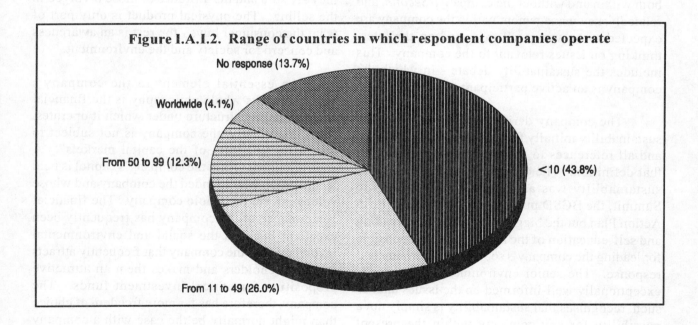

Figure I.A.I.2. Range of countries in which respondent companies operate

Annex II

THE CASE STUDIES

A. Case 1 (UK)[35]

Case-1 plc is a high-profile consumer-products company specialising in cosmetics, health and skin care products. Manufacturing takes place in the UK and in the USA with sourcing and distribution worldwide. An explicitly "values-centred" company which identifies social and environmental values in its mission statement, memorandum and articles, Case-1 plc maintains a high profile by campaigning on both social and environmental issues. This has two major influences relevant to this report. *First*, the company operates a high degree of transparency both within and without the company. *Second*, and relatedly, key management within the company are expected to be very well-informed on the latest thinking on issues relevant to the company. This includes the sustainability debate - in which the company is an active participant.

The company derives its understanding of sustainability initially from the Brundtland Report and all references to sustainability are related to that definition. The company's understanding of sustainability was also influenced by the Rio Summit, the BCSD publications and the EU Fifth Action Plan but the biggest influence is the reading and self-education of the management responsible for leading the company's social and environmental response. The senior environmental manager is exceptionally well-informed on the issues and, as such, recognises that sustainability is simply not a possibility for any company within the present economic system.

The quest for economic growth, as demanded by national and international financial institutions, is the cause of much environmental and human exploitation.

It does not pay to be sustainable. Good housekeeping saves money but the pursuit of sustainability is beyond good housekeeping - and can cost.

The company's view of sustainability has been developing. Now recognising that sustainability, as such, is difficult to imagine, let alone achieve, the issue is the path towards sustainability. This involves looking for more humanly-sensitive, more environmentally-sensitive and less un-sustainable activities throughout the whole life-cycle of the firm and its products. This has led to more human-centred approaches to product procurement (with especial regard for indigenous peoples), a range of initiatives almost unique in business on energy replacement and waste treatment and high rhetoric of concern for, not just the quality of the products, but the way in which they are sold and the attitudes of those involved in the selling. The physical product is only part of what the consumer buys - the rest is an awareness and concern for society and the environment.

An essential element in the company's approach to its whole philosophy is the financial and accounting structure under which it operates. Most importantly, the company is not subject to the "discipline (sic) of the capital markets". A controlling interest in the company's capital is held by the group who founded the company and whose values inform the whole company. The financial performance of the company has frequently been excellent but it is the social and environmental involvement of the company that frequently attracts other shareholders and makes them an attractive proposition for ethical investment funds. The company, therefore has far more freedom of choice than might normally be the case with a company more exposed to financial market choice.

Relatedly, the financial accounting system is (in their terms) "minimalist". It serves the essential functions of plotting cash flows, paying wages, logging debtors and creditors and constructing the annual financial statements but, beyond this, the accounting and financial system is relatively subdued. Accounting and financial controls and performance evaluation are generally less important than other performance criteria. This is especially so when social and environmental issues are under consideration.

The company considers that the accounting and financial orientated concern over (a) financial accounting issues such as contaminated land, provisions, contingent liabilities and financial auditing implications; and (b) changing the accounting system to be more responsive to environmental needs on issues such as capital budgeting, waste, energy, emissions and so on, are largely irrelevant to them. This is *not* to say that the underlying issues are not critical - they are - but rather this company will not see these issues through the lens of accounting. It prefers to address such issues directly. Accounting is thus the last place such an enlightened company would look for devising a path towards sustainability.

The company is also sceptical about using accounting as a method of either reporting externally or trying to measure and capture sustainability. There are other, more appropriate, ways of doing this and so why bother trying to measure sustainability through the accounting system when the issues can be addressed in a more direct manner.

The company *does* recognise, however, that: (a) every company must seek to educate its financial market stakeholders - but would rather do this through other non-financial information; and (b) for the more accounting and finance orientated organisation, it may prove practically and politically necessary to employ accounting as the means for change. The company remains sceptical about this approach though.

B. Case 2 (NETHERLANDS)[36]

Case-2 NV is the foods division of a multinational company operating, principally, in the foods, detergents, personal products and speciality chemicals industries. Although a well-known company through its consumer products, Case-2 NV and the group of which it is a part has not maintained a high profile on environmental issues:

The company has not generally made a lot of fuss about its outside image. We may be moving towards environmental reporting in the future - but what should be reported? How should we do environmental reporting? What reliable and useful information is needed which can be collected. There is a major potential conflict between the environmental information needs of management and the environmental information wants of external stakeholders.

By contrast, the company does have a fairly sophisticated and systematic approach to reporting to its employees. The company has signed "environmental charters" and the principal effect of this has been to give leverage to the environmental departments of the company to take environmental issues forward throughout the group.

The company's principal mechanism for addressing environmental and sustainability issues is a two step process. The initial conceptual step has been careful consideration of the concept of *ecological space*. The planet only has a limited ecological space for human activities, so how much of that space is available for foods? This leads to a recognition that looking at refining certain products to reduce their environmental impact or substituting products, processes and raw materials in order to reduce an organisation's use of environmental space really achieves very little globally. Each choice involves trade-offs - more organic foods involves greater land use, more health-conscious products often involve increased energy and processing, one company's raw material or energy selection may simply pass on additional environmental impacts to other parts of the activities' life-cycle. The key driver of use of the ecological space is *lifestyle*. This is the responsibility of people as a whole, not just this company.

The second element is the development of a sophisticated and detailed Life-Cycle Assessment of the group's products and activities. Every step in the whole life-cycle is assessed, costs are attached, resource and energy usages identified, and alternatives considered together with the costs of those alternatives.

LCA is not a threat. It will help a very great deal in addressing environmental issues and helps a lot in explaining to other parties what the company actually is doing. It shows just how much is not entirely up to the company. There is a lot of good environmental practice in the company. The LCA shows that simple solutions are often more complex and involve other trade offs. A major issue is that a better environmental choice will have both financial costs and environmental costs attached to it.

This leads onto the conclusion that sustainability needs a dramatic change in lifestyle:

We all would like to believe that sustainability is achievable without a major change in lifestyle. Unless there is a major change in our attitudes to our way of life there will be no real improvement in our environmental impacts. This arises from the LCA analysis - it is a natural conclusion.

Business would not necessarily come up with the same arguments about the difficulty of sustainability because they have not had the opportunity to consider it in as much detail as I have.

A company can only go to the point where pollution prevention pays. It is unrealistic to expect it go further. Every time a company introduces a new product, the efficiency of the company declines whilst it gets the volume back up. Companies must introduce new products because the market demands them - demands the continuous improvement to which the consumer is committed. There is an environmental cost attached to that though. New product decisions will always dominate in environmental terms because every company is performance (financial) driven:

Financial criteria will always dominate the company. It is my job to turn the environmental elements into money - at least to show the effects

of ignoring environmental issues. Bad environmental decisions now will have financial effects in the future.

Environmentally-related costs are identified by the accountants and are put into the LCA. People have a healthy personal interest in the environment in this company and are very cooperative *until* we have to deliver the profit. Company performance must be the dominant concern. Until consumers change their lifestyles and institutional investors are prepared to accept a lower dividend in return for lower acid rain, there is not much we can do.

Accounting has a function within this framework. Currently the company charges back environmentally-related costs to line managers. The accountants are seeking to identify the cost drivers (e.g. design) whilst the environmental department is seeking to identify the environmental drivers - which economic decisions drive environmental impact.

Sustainability is not yet possible for companies. Accounting for sustainability? It may have possibilities but it would be wrong to move on these issues (e.g. The EU Fifth Action Programme) or make recommendations until the practicalities and implications have been fully worked out.

D. Case 3 (ITALY)[37]

Case-3 SpA is a large company within the Italian context producing crackers and intermediates for the chemical industry, plastic products and materials, fibres, elastomers, detergent intermediates, fine and speciality chemicals. The company has operations in over 20 countries worldwide with a primarily European focus. While presently not quoted on the stock exchange the company has a strong commercial orientation and could expect to become privately owned in the future. Case-3 SpA is leading environmental reporting in Italy (the environmental report having recently won a national level award) which appears to be driven by the realisation that the credibility and continued operation of the company requires *"concrete initiatives whose effects are visible and, wherever possible, quantifiable"*. In recent years the licence of this company to operate has been called into question - they have had major environmentally-related problems and this, combined with public ownership, has put considerable pressure on them to 'get their house in order'. There is also consideration of extending the environmental auditing and reporting undertaken to incorporate social aspects such as the employment and community impact of their operations.

The four contact points within the company approached sustainability from a variety of points. While there was awareness (with one notable exception) of the Brundtland Report, Agenda 21 and the EU's Fifth Action Plan, no one could provide an explicit definition of sustainability, rather, examples of what could be considered to be moves towards sustainability were offered. These included: changing processes to reduce the amount of wastes generated; substituting less toxic by-products for more toxic by-products; improving energy and raw material efficiency in manufacturing; and the provision of new products and processes to satisfy customer demands. Given the complexity of the ideas behind sustainability and the need for fundamental societal change in response to the concept, this company clearly has only begun the process of understanding the concept.

The above understandings can be contrasted with the company's accountants who had no knowledge or comprehension of the term and who exhibited an "accounting-constrained" attitude towards their role in assisting the company come to grips with the environmental agenda. They saw their role as being limited to being as flexible as possible to allow the environmental department's questions to be answered, but otherwise felt existing accounting regulations provided adequate scope for dealing with environmental matters.

The company's position on the barriers to sustainability and mechanisms that will move them towards sustainability were uncompromisingly expressed thus:

Most chemical companies are owned by oil companies which has slowed down investment. If petro-chemical companies were not owned by the oil companies they would go back to raising finance on the capital markets rather than using in-house finance. This would be better as the proper relationship between the companies and the capital markets could be established. The capital markets will not favour projects with environmental risks ... therefore environmental improvements would be supported by the capital markets.

People need to get back to the old religion of making money and risking things. If industry went back to risking things sustainable development would happen.

Basically what is needed is to substitute better products for existing ones, increase the efficiency with which we do business and de-toxify what we have at present while maintaining economic growth and everything will be OK. This should happen in the next cycle of investment ... say 10-20 years.

These views appear to be at odds with the bulk of the literature in this area (with the exception of publications of the Business Council for Sustainable Development) and involve an enormous faith in market mechanisms. This is also in direct contrast with one interviewee's **personal** views on the matter where continuous growth was questioned and the intractability and importance of intra-generational equity was noted.

What makes this company interesting within the context of accounting is the existence of an environmental information database which has been developed in conjunction with an external, independent research organisation. The approach, in essence, is motivated by an attempt to mirror, at a micro-level, the environmental accounting attempts being made at the national level in Italy. Based, loosely, on a mass-balance perspective of the organisation the developing "*environmental balance sheet*" (as it is called) attempts to identify all inputs, outputs, emissions and leakages and then attach financial numbers to these flows. In broad terms, the "environmental balance sheet" should allow an identification of where current environmental costs are falling and what cost-benefit trade-off might be possible from changes in processes within the company or from changes to the inputs to those processes.

This, widely-publicised, experiment certainly has focused the minds of management on the linkages between the environmental and economic aspects of the company. It also seems that its basis in an oköbilanz framework will, in time, provide a useful "management tool". The approach is still experimental and work continues to refine the method and deal with its inconsistencies and problems.

Case-3 SpA are perhaps representative of companies in which there are varieties of views and perceptions of the role of environmental issues within an economic unit like a company. A well-informed environmental management team find themselves in tension with a *relatively* well-informed senior executive team whose understanding of sustainability is somewhat primitive and, to whom, environmental issues must always be placed within a "wider" context of the economic success of the business.

D. Case 4 (SWITZERLAND)[38]

Case-4 SA, operating in Western Europe, the Middle East, Africa and Eastern Europe, is primarily concerned with the manufacturing and selling of chemicals and plastics. There is a high level recognition of the concept of sustainability within the organisation and this is reflected in their public statements. The recognition of sustainability as a concept of importance arises from concern about the likely competitive environment the company is likely to face and the effect of government regulation on their operations.

The company draws on a variety of sources for its understanding of sustainability including environmental consultants, the work of DG XI, the European Environmental Bureau (which represents

some 150 environmental organisations in Brussels) and they have been actively supporting work by the Business Council for Sustainable Development. This company's approach to the area is typical of many transnational corporations and can be expressed thus:

Governments need to set clear, consistent, tax neutral and common sense targets for environmental performance and then give business the freedom to innovate and deliver the desired performance. This will lead to sustainability within a time frame of approximately 10-30 years ... ultimately there will be a new generation of products that will build a sustainable future.

One of the striking features of this approach is the confidence with which individuals and companies are able to determine the "best" approach to solving complex environmental, economic and social problems.

The company was able to express how they saw the role of economic growth in the pursuit of sustainability. In the future it was expected that society would be "de-materialised" in that much less raw materials and energy would be required to produce products. This process will inevitably lead to a decrease in the volume of production, however, economic growth will be sustained with a change in business activity towards incorporating a greater service component to physical production. For example, a chemical manufacturer is likely to be involved in recovering their product from customers, cleaning the contaminated material, disposing of waste residues and re-supplying the product back to customers. This substantially *closed system cycle* will generate extra economic growth. In this way the environment's capacity to assimilate pollution, which is currently not priced, will be replaced will economic activity. This has been described as a qualitative change in the nature of economic growth. It is likely that such a change will reduce the current level of un-sustainability but it is very difficult to know if this change will be sufficient.

The role of accounting and finance was highlighted as being important for the company's move towards more sustainable operations. The capital markets and financial institutions were identified as presently inhibiting sustainability, however, their role was considered to be changing. In particular, the need to develop some kind of full cost accounting was highlighted. Full cost accounting was thought to place an environmental price on the materials used and sold in a way that reflected the full environmental cost of the product. The interviewee felt that this mechanism would only work if the price was passed on to the consumer because that was where control over consumption of resources and pollution rested. There was much resistance to the idea of applying taxes or costs at the input level because it was felt that this would not result in changes in consumer behaviour. The detailed mechanisms to full cost accounting were not at all developed.

This company appears to reflect a mainstream view of the possibilities of the pursuit of sustainability by business and the relative roles and responsibilities of government and business. It also reflects an uncritical acceptance of the current structure of business operations and the ability of current structures to deliver the fundamental changes that are needed for a sustainable future.

E. Case 5 (AUSTRIA)[39]

Case-5 AG is a privately owned company with a significant presence in the Austrian economy. The enterprise manufactures telephones and has a licence to produce telephone switching exchanges using the know-how of a large transnational corporation. They operate in a highly competitive industry subject to rapid technological innovation. Their operations, however, are relatively "clean".

Concern for the environment developed from both a top down (there is significant personal concern at senior levels) and bottom up approach (details of initiatives come from employees - there are approximately 100 employees organised in 14 project teams) and is manifest in a programme of continuous environmental improvement. As part of the pursuit of more environmentally friendly operations interviewees had come across the concept of sustainability which is seen as being *"more than a concern for the environment"* and is

likely to involve *"more than environmental management"*.

While not being aware of the sustainability "touchstones" that were common in other interviews (the Brundtland Report, EU's Fifth Action Plan or the Business Council for Sustainable Development) individuals displayed a reasonably sophisticated understanding of sustainability.

It is a question of our responsibilities as a company and as human beings ... while there are legal duties there is also a moral duty, particularly as we live in a very privileged part of the world.

We are definitely not sustainable at present ... I would have thought that we were more sustainable 20/30 years ago and have spent recent years moving away from it fast.

We have to ask how can we, as human beings in the twentieth and twenty-first century change our way of life ... continuous growth in material goods and consumption is not possible.

Europe does not need to grow but there is an issue with lesser developed countries ... they should not be satisfied with what they have and we cannot expect them to be so ... where and how they develop is their decision not ours ... we do not have the right to dictate to them how they should develop.

These statements seem to reflect a values-driven base from which social and environmental issues could be addressed. This is not the predominant values of the organisation (which are strictly financial) but may be indicative of their size and sense of control over the organisation.

Interviewees identified fundamental barriers to sustainability within their industry. They postulated that sustainable development could be enhanced by the manufacture of durable goods that are designed for a long life. Indeed items that they had manufactured as long ago as 1905 could still be used (however, fashion and other factors dictated that they were not acceptable to customers). This perceived need for durable goods was contrasted with the fact that within the electronics industry the manufacture of many components has become so inexpensive that durability had ceased to be an issue. In addition, rapid technological innovation makes items obsolete even though they have a long useful life. These twin, industry-specific, pressures illustrate some of the complexities that any business moving towards sustainability will need to address. They equally illustrate the problem of asserting that the will of consumers and the rigour of competitive pressures can unproblematically lead to sustainability.

In accounting terms this company has produced an input/output eco-balance since 1991. This is constructed from information systems and has been used to see how components in production change over time, along with percentages of energy and wastes produced. This is used as a starting point in identifying which parts of the operations to examine further in the hope of finding both financial and environmental improvements. Unsurprisingly, the thrust of environmental management has been to seek competitive advantage and to generate financial savings. This has resulted in changes in the production processes and the development of technology specifically designed to clean and recycle chemicals used in production. The use of input/output analysis in both physical and financial terms combined with life cycle assessment methodology (which is not specifically undertaken in this company) was seen as a potential way forward in assessing an organisations degree of unsustainability and as a guide to future actions.

F. Case 6 (SPAIN)[40]

Case-6 SA is one of Spain's largest companies and generates and supplies a large proportion of Spain's electricity - from nuclear-, fossil-fuel-, hydro- and wind-powered generators. The company is not a TNC[41] but, given its size, the importance of energy generation and supply to the sustainability debate and the contrasting perspectives and initiatives of the company, its inclusion in the casestudies seemed highly apposite.

In one major respect, Case-6 SA may be thought of as a "typical" company[42] (rather than as a path-breaking one) on the issue of sustainability. Its general view is that sustainability

is both too theoretical and too long term a concept to be of immediate and pressing concern to the company. Thus, because the issue cannot be adequately defined and the future (in every respect) is now so uncertain, it is neither possible nor appropriate for the company to address sustainability directly. No effort has been expended, therefore, in attempts to define and assess what sustainability means for the company. More particularly, the current incentives lean towards encouraging energy consumption - not reducing it.

However the principal constraint is one of technology. Simply stated, the technology does not exist which offers serious alternatives to current practice. There *are* trends which may be leading towards more sustainable options, however. For example, technology is leading energy generation towards the building of *smaller* rather than enormous plants. This not only leaves a company with a greater flexibility (on the portfolio theory effect) but also reduces the centralising tendency of power generation and reduces the distribution networks - with the associated reduction in power losses.

Case-6 SA is a major contributor to the quality of life through the provision of reliable electricity via industry best practices. Quality of life is a factor in sustainability which a single company cannot easily address on its own. The path towards sustainability will be iterative and pluralistic in which all individuals and groups make decisions and contributions. All have different contributions to make. Coordination of these different contributions, however, must be done at the pan-national level by (for example) the EU or the United Nations.

The company's principal contribution to the path to sustainability is therefore through, in effect, an EMAS approach. By seeking maximum efficiency, developing environmental management systems and encouraging (and monitoring) the latest technological developments, Case-6 SA can continue to supply the demand for a (currently) essential element of the Spanish way of life as efficiently and effectively as possible.

But to portray the company as uninformed and unadventurous would be incorrect. The key personnel in the company have a significant knowledge of the leading sources of influence on the global sustainability debate but they recognise - probably correctly - that few people (if any) really understand the implications of (for example) full cost pricing for business.

The company's major innovation is its own approach to sustainability - given its basic belief about the role of business. Stressing that the long-run outcome is not a predictable function of current factors and present trends, Case-6 SA see the root of the problem lying in conventional economic thinking and its application through accounting. Recognising the (now well-established) problems deriving from the non-pricing of ecological and social factors, the company is trying to find ways to develop their internal accounting system to increasingly recognise environmental factors. This is not yet well-developed although a project is underway which is beginning to identify environmental spending and to feed the environmental into all aspects of the company's business. This will then form the basis for a more developed experiment in environmental accounting. What is so unusual about this is that the company is not constrained (as yet) by capital markets or profit considerations.

Typically, however, the environmental department and the accounting function are meeting conflicts. Whilst such matters as environmental improvements and capital investments are fully integrated, there is still a reluctance in the company to either account for contingent liabilities or provisions - or to release cash for land remediation when there is no foreseeable cash benefits to doing so.

This leads back to the need for firm guidance at national and international levels. Spanish law shows little sign of forcing companies to internalise current externalities (for example contaminated land) and solutions, such as environmental accounting, are in their infancy. The company would be happy to see tax structures changed to force the internalisation of the environment into prices and is actively encouraging exploration of sustainability accounting. However, the company

stresses that suggestions in either area must be well-thought through and fully researched. There is too much evidence of quick fixes, policy-on-the-run and sustainable suggestions based on too-thin a foundation of knowledge and understanding. This is not helpful and actually alienates those businesses which *are* supportive of such initiatives.

Energy is probably the number one environmental issue but it is demand driven in a biased market place. Prices favour consumption - not conservation - and encourage the consumption of fossil fuels. This must change if progress is to be made. Such changes should be carefully linked to other changes in the tax system and to the development of a realistic environmental accounting.

Annex III

THE QUESTIONNAIRE

SUSTAINABLE DEVELOPMENT AND ACCOUNTING:

Incentives and Disincentives for the adoption of sustainability by Transnational Corporations

A RESEARCH INVESTIGATION

PLEASE MAY WE HAVE SOME PERSONAL DETAIL ABOUT YOU AND YOUR ORGANIZATION?

ALL ANSWERS WILL BE TREATED WITH TOTAL CONFIDENTIALITY - WE WOULD NOT RELEASE ANY INFORMATION WITHOUT YOUR PRIOR APPROVAL
Organisation's Name:
Name of person completing the questionnaire:
Position of person completing the questionnaire:
Country where questionnaire is being completed:
Number of countries in which the organisation operates:

PART I:

The first part of this questionnaire explores the introduction of the term "sustainable development" into the arena of business. In particular, we are trying to find out what this term may mean at an organizational level.

PLEASE NOTE: We make no distinction between "sustainability" and "sustainable development". If you wish to do so please make this explicit.

The following are views are taken from a variety of sources, concerning business's role in a sustainable future. Please indicate the extent to which you agree or disagree with each statement **by circling one number for each question.**

Business' pursuit of sustainable development.....	Strongly Agree				Strongly Disagree
..is impossible	5	4	3	2	1
..requires a partnership approach from government, business and society	5	4	3	2	1
..involves fundamental changes in attitudes and values	5	4	3	2	1
..is likely to be complex and require extensive change	5	4	3	2	1
..is achieved in my organisation	5	4	3	2	1
..is compatible with the pursuit of economic growth	5	4	3	2	1
..will cost significant amounts of money	5	4	3	2	1
..requires complete transparency and accountability	5	4	3	2	1
..will involve all parts of society in discussion and implementation	5	4	3	2	1
..is meaningless - we can only talk of decreasing our *unsustainability*	5	4	3	2	1
.. is meaningless until there is sufficient consensus on what it means for business	5	4	3	2	1
..is impossible with our current accounting systems	5	4	3	2	1
..is essentially a concern with the needs of future generations	5	4	3	2	1
..implies a concern for the current distribution of wealth between nations	5	4	3	2	1
.. is incompatible with the profit ethic	5	4	3	2	1
..means tackling both **social** and environmental problems	5	4	3	2	1
..is synonymous with the development of environmental management systems	5	4	3	2	1
..requires balancing the needs of the economy with environmental protection	5	4	3	2	1
..requires the support of the financial markets	5	4	3	2	1
..means considering third world peoples, economies and resource bases	5	4	3	2	1

Please indicate the extent to which the following documents and organisations have influenced your understanding of sustainable development.

	Important influence			Did not influence	✔ = never heard of this source
The World Conservation Strategy	5	4	3	2	1
Our Common Future (better known as the Brundtland Report)	5	4	3	2	1
Agenda 21 and the other documents produced in Rio at the Earth Summit	5	4	3	2	1
The International Chamber of Commerce and their Business Charter for Sustainable Development	5	4	3	2	1
Publications from the Business Council for Sustainable Development	5	4	3	2	1
The European Union's Fifth Action Plan - "Towards Sustainability"	5	4	3	2	1
The media - e.g newspapers, magazines, TV	5	4	3	2	1
Environmental pressure groups (such as Greenpeace - please specify)	5	4	3	2	1
Your own company/organisation	5	4	3	2	1
Books and/or economic journal aricles & papers	5	4	3	2	1
Documents published by your national government	5	4	3	2	1
Documents/papers produced by your industry association or professional body	5	4	3	2	1
Personal membership of non-governmental organisations (please specify)	5	4	3	2	1
Other (please specify)	5	4	3	2	1

Is an exact definition of sustainable development necessary?

Yes	No
1	2

PART II:

This part of the questionnaire attempts to explore the implications that the pursuit of sustainable development would have for <u>your organization</u>. Further, it attempts to look at how moves towards sustainability could be made.

Yes	If 'yes' could you please outline where
No	If 'no' could you please outline the extent to which sustainable development is informally recognized by your organization

Please indicate the degree to which the following have encouraged/discouraged <u>your company</u> in moving toward sustainability.

Moves towards sustainability	Strongly encouraged			no effect			Strongly **discouraged**
The United Nations	7	6	5	4	3	2	1
The European Union	7	6	5	4	3	2	1
Your National Government	7	6	5	4	3	2	1
Industry Associations (please specify)	7	6	5	4	3	2	1
Your customers	7	6	5	4	3	2	1
Your suppliers	7	6	5	4	3	2	1
The capital markets	7	6	5	4	3	2	1
Law - actual and impending	7	6	5	4	3	2	1
Market opportunities	7	6	5	4	3	2	1
The financial accounting system	7	6	5	4	3	2	1
Financial institutions	7	6	5	4	3	2	1
Your employees	7	6	5	4	3	2	1
Pressure groups (please specify)	7	6	5	4	3	2	1
Other parts of your organisation (please specify)	7	6	5	4	3	2	1
Non-governmental organisations	7	6	5	4	3	2	1
Other (please specify)	7	6	5	4	3	2	1

The next table outlines a number of activities that may relate to the pursuit of sustainable development. For each area could you please indicate the extent to which you believe that each activity will move <u>your organization</u> on the path towards sustainability and indicate whether or not your organization is active in these areas.

	Will contribute towards sustainability		Will **not** contribute towards sustainability		✔ = doing something x = not active ? = not sure how to
Reduction in the consumption of resources	5	4	3	2	1
Reduction in energy usage	5	4	3	2	1
Reduction in the volume and toxicity of wastes	5	4	3	2	1
Product design to increase the life of the product and improve repairability	5	4	3	2	1
Recycling	5	4	3	2	1
Commitment to an environmental management and audit system	5	4	3	2	1
Policy statements regarding the pursuit of sustainable development	5	4	3	2	1
Targets for moving towards sustainability	5	4	3	2	1
Reporting on the achievement of sustainable development targets	5	4	3	2	1
Use of social and environmental criteria in capital budgeting and investment decisions	5	4	3	2	1
Being open to scrutiny by groups and individuals outside of the organisation	5	4	3	2	1
Redesigning the accounting system - e.g. for full cost accounting or for environmental provisions	5	4	3	2	1
Providing information to shareholders and financial community on your social and environmental impact	5	4	3	2	1
Other (please specify)	5	4	3	2	1

Please indicate how important the following constraints are in <u>your organization's</u> ability to pursue sustainable development.

	Very important constraint				Not an important constraint
The definition of sustainable development is too unclear - hence unable to know what an organisation should do	5	4	3	2	1
How this organisation could move towards sustainability is unclear	5	4	3	2	1
Unsure of the organisation's role in pursuing sustainable development	5	4	3	2	1
There is inadequate government guidance on how to pursue sustainable development	5	4	3	2	1
There is inadequate government support to enable the organisation to pursue sustainable development	5	4	3	2	1
The pursuit of sustainable development is not important for this organisation	5	4	3	2	1
There are too many other pressing issues demanding attention for the organisation to concentrate on sustainable development	5	4	3	2	1
Pursuing sustainable development will cost the organisation too much	5	4	3	2	1
There is no demand for the organisation to pursue sustainable development	5	4	3	2	1
The current measures of success do not recognize sustainable development	5	4	3	2	1
Our current accounting systems do not reflect or encourage progress towards sustainable development	5	4	3	2	1
Other(s) (please specify)					
....	5	4	3	2	1
....	5	4	3	2	1
....	5	4	3	2	1
....	5	4	3	2	1

PART III:

This final section of the questionnaire attempts to discover what role accounting, finance and reporting play - and could play - in <u>your organization's</u> progress towards sustainability.

Organization vary considerably in the emphasis that they place on accounting systems and accounting/financial criteria. We would welcome an indication of the extent to which accounting and financial criteria are important in making decisions within your organization.

Accounting and finance systems and criteria....................	..dominate organizational decisions		.. are just one set of criteria	..are dominated by other criteria	
Corporate strategy	5	4	3	2	1
Medium and short term planning	5	4	3	2	1
Divisional performance evaluation	5	4	3	2	1
Managerial performance evaluation	5	4	3	2	1
Remuneration of senior executives	5	4	3	2	1
Capital expenditure and investment	5	4	3	2	1
Make or buy decisions	5	4	3	2	1
Relationships with the financial community	5	4	3	2	1
Staff training	5	4	3	2	1
Research and development initiatives	5	4	3	2	1
Community involvement	5	4	3	2	1
Environmental projects	5	4	3	2	1
Environmental standards and targets	5	4	3	2	1
Purchasing decisions	5	4	3	2	1

Please indicate how familiar are you with the following accounting techniques and whether your organization is using them.

	Very familiar			Un-familiar	✔ = we use this
Full cost accounting (e.g.the Tellus Institute)	5	4	3	2	1
Full cost pricing (e.g EU Fifth Action Plan)	5	4	3	2	1
Sustainable cost calculation (e.g. CSEAR)	5	4	3	2	1
Net environmental value added calculations (such as that done by BSO/Origin)	5	4	3	2	1
Extended cost benefit analysis	5	4	3	2	1
Life Cycle Assessment with financial numbers assigned to physical flows	5	4	3	2	1
Social Auditing (e.g Traidcraft or Sbn Bank)	5	4	3	2	1
Eco-balance sheets/Oköbilanz (e.g the Danish Steel Works) with financial numbers assigned	5	4	3	2	1

Please indicate how important the following accounting-related activities are in your organization's move towards sustainable development.

	Very important				Not at all important	✔ = doing something x = not active ? = not sure how to
External environmental reporting	5	4	3	2	1	
Liaison with the financial auditor	5	4	3	2	1	
Creating environmental provisions	5	4	3	2	1	
Environmental contingent liabilities	5	4	3	2	1	
Disclosure in Financial Statements	5	4	3	2	1	
Environmental and ethical criteria in the investment process	5	4	3	2	1	
Preparation of Eco-balance sheets	5	4	3	2	1	
Financial estimates of externalities	5	4	3	2	1	
Identification of the (socially & environmentally-related) costs of:						
(a) wastes and waste disposal	5	4	3	2	1	
(b) energy usage and waste	5	4	3	2	1	
(c) packaging retrieval and disposal	5	4	3	2	1	
(d) regulatory compliance	5	4	3	2	1	
(e) planning consents	5	4	3	2	1	
(f) breaking regulations	5	4	3	2	1	
(g) negotiating with regulators and lawyers	5	4	3	2	1	
(h) environmental management and audit schemes	5	4	3	2	1	
(i) environmental investment	5	4	3	2	1	
(j) community involvement	5	4	3	2	1	
(h) environmental product design	5	4	3	2	1	
Full cost accounting	5	4	3	2	1	
Allocation of environmental overheads to products and processes	5	4	3	2	1	
Relationships with shareholders and the financial community	5	4	3	2	1	
Meeting with analysts and ethical investment trusts	5	4	3	2	1	
Financial risk assessment	5	4	3	2	1	
Insurance	5	4	3	2	1	
Asset revaluation of:						
(a) land	5	4	3	2	1	
(b) fixed assets	5	4	3	2	1	
(c) inventory	5	4	3	2	1	
(d) other (please specify)	5	4	3	2	1	
Assessment of market opportunities	5	4	3	2	1	

Chapter II

THE INTEGRATION OF ENVIRONMENTAL PERFORMANCE INDICATORS WITH FINANCIAL INFORMATION BY TRANSNATIONAL CORPORATIONS

Report by the UNCTAD secretariat

Summary

This report is on a study that was undertaken to examine the potential for integrating financial and environmental performance reporting. The analysis is supported by case studies of six Swedish enterprises. It has been concluded that the integration of business and environmental performance reporting is likely to develop for several reasons. It will probably not, however, come about as add-on information to traditional enterprise reports. The potential is to be found in refining the format and structure of performance reporting in general. Enterprises with clear and understood site dependency, low exposure to short-term pressures from the financial markets, high integrity in management, materially viable production processes, and quality-conscious customers will be early in developing new performance measures. Reporting practices which support internal control in enterprises are a sound basis for the development of information disclosures. Agencies within governments should be prepared to differentiate their policies regarding transnational corporations. However, the conditions for sustainable development differ among enterprises, industries, countries, and regions of the world. Governmental activities supporting a sustainable development should differ accordingly.

INTRODUCTION

The Intergovernmental Working Group of Experts on International Standards of Accounting and Reporting (ISAR) has previously investigated the disclosure of environmentally relevant information by transnational corporations. The findings so far have indicated a general impression of sluggishness of concerns about environmental issues in the business agenda. With this background, ISAR decided to study the subject of environmental disclosures by enterprises from another approach: the objective was not to obtain a quantitative record of certain corporate practices but to use a few case studies as a point of departure for a qualitative analysis of potentials in the field. The main focus in this report is ways to integrate environmental performance indicators into a framework of general business performance measures.

It is initially necessary to elaborate on the context of the issue. Who are the users of company reports? What reasons would justify an integration between financial and environmental reporting? To what extent should internal control considerations influence standards for external disclosures (and the other way around)? Generally, environmental performance indicators may be used for several reasons such as an assessment tool within internal environmental management systems, a way to inform potential investors and shareholders, a media for dialogue and negotiations with authorities and with suppliers and customers, etc. Still it is an open question about whether that kind of information should be integrated into regular company reports, if it should be integrated in some other way or if it is better if the information is reported separately in some way.

The financial performance of transnational corporations (TNCs) is important to financial markets world wide. Participants in those markets tend to treat corporations in a standardized way which is one reason behind claims for rigor in company reports; people with no experiences with the company as a physical entity whatsoever should get a proper impression of it.

Environmental performance is traditionally not in the set of relevant information for the investing community. Until recently environmental performance was only a local issue: smoke from the chimney, chemicals in sewage water and noise from traffic. Politically, the issues have been important for a long time at the municipality level, while international regulation of these activities has occurred only lately. Neither the local authorities nor international bodies have had any specific use for the integration of business and environmental performance.

Many persons consider that the use of integrated performance measures should be part of internal control systems in the first place. Responsive enterprises with a clear strategic view of their role in world-wide throughput of matter and energy, with policies for ethical sales, climate change, local responsibilities, throughput thrift, etc., will most certainly need reports on performance covering all such aspects. Then integration is important.

After the world conference on the environment in Brazil in 1992, the idea of sustainable development as the first priority in all political assemblies and as the prime guideline for individual and enterprise actions became generally accepted, at least in principle. One practical consequence is a need for responsive TNCs. Governmental and international bodies must then form a supporting environment for such enterprises.

Disseminating the experiences of ways to integrate performance reporting may both facilitate access to "responsive" practices and, indirectly, raise the general expectation for enterprises to be responsive. If this is successful, it is likely to lend support for international standards for environmental regulations and taxation. That will, alongside explicit standards for performance information disclosures, give responsive enterprises a competitive advantage. It will also give readers of advanced performance reports solid guidelines for how to interpret the facts and figures.

I. THEORETICAL FRAMEWORK

It is often assumed that financial performance is generally considered to be of primary interest and environmental performance tends to get isolated and downgraded if it is expressed in a mode very different from generally used economic language. The study will now consider the accuracy of that assumption and ways to integrate environmental and financial considerations in enterprise reports.

In the literature there are two broad classifications for management behavior: one can be recognized as "empirical" and the other as "normative". The empirical classification is based on propositions of the following types (Andrews 1949, Cyert & March 1963, Chandler 1977, Earl 1983, Douglas 1986, Drucker 1993 and others):

(a) Companies try to keep their risks low. Business management is seldom gamesmanship. When risks are taken they are out of perceived necessity.

(b) Companies aim at long term survival. One should expect special reasons when shortsightedness prevails in management thinking.

(c) Companies behave "ecologically" in relation to their environments; they try to adapt to environmental changes and they try to make the environment adapt to the company.

(d) Leadership in companies is idealistic. Political and technical ideas, personal dreams and cultural blindness form strategies and company structures. Cost-and-revenue arguments normally appear late in decision processes.

(e) Decision processes in companies are most often characterized by limited rationality. The idea of rationality is almost never abandoned, but relatively few alternatives are seriously considered when making decisions and all criteria are seldom, if ever, weighted against each other.

The empirical model generally provides a favorable picture of company behaviour. These features can be expected to be found in healthy and prosperous companies where management has considerable integrity. That integrity may get challenged in various situations. One such challenge occurs when the company in question is a subsidiary of a transnational group and the parent company management prescribes specific behavior and performance at the subsidiary level.

Functional differentiation within a group may also be reflected in a distorted view of the real conditions of the operations in each local company. A company which is eager to attract money from the financial markets (whether it is debt or equity money) becomes dependent on the speculative short-sightedness that guides those markets. Financial operations provide a high opportunity cost for the internal use of money and may thus influence corporate structures. Generally, when dealing with economic hardships, managements have to act with less long-term considerations. Then the empirical evidence comes close to the standard assumptions which direct the normative model.

The normative model prescribes company characteristics such as price orientation, reactionary rather than innovative behavior, short-term profit maximization, and a minimum legal compliance attitude to environmental issues.

When debating company roles in achieving sustainable development, the normative model dominates the scene. Consequently, the focal issue most often is about forcing companies considered as being extremely short-term oriented to disclose environmental and long-term information.

This study acknowledges this background, but still gives the empirical model a chance by avoiding descriptive techniques which put the environment outside business considerations at the outset. Each enterprise in the case studies is described as a value-driven set of capital and operations, which are defined as the *corporate effort*. *Values* in this respect may be recognized as specific service qualities strived for, shareholder claims or other strategic targets as they are articulated by corporate management. In this way the environment appears on equal footing with other resources which are utilized and/or assumed: the specific set of capital and operations under management requires a throughput of human, financial and material resources. Figure II.1 presents a formal picture of the basic scheme in the case studies.

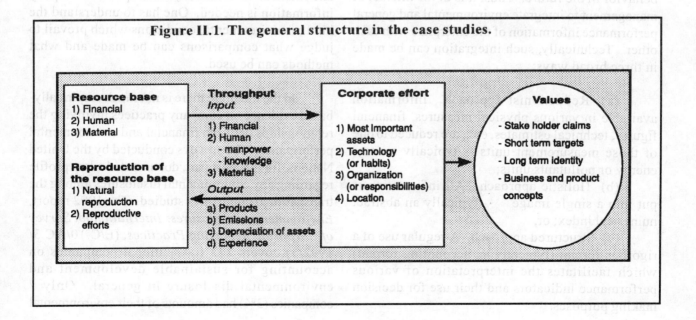

Figure II.1. The general structure in the case studies.

II. SUSTAINABLE DEVELOPMENT AS A MANAGERIAL ISSUE

How does the environment appear as a managerial issue? An enterprise may reduce air emissions by a certain percentage or volume or improve the quality of its emissions. However, if this enhancement is at a significant cost and has consequently led to an increase in investment, how should this be best reported? This question has an obvious time dimension since the investment in a long-run context may both save costs and increase revenues, while it may result in a loss if forced on an enterprise with short notice. Alternatively, energy consumption may be reduced through a more fuel-efficient industrial process which may also result in cost savings in the long run. Another example could be more stringent pollution regulations which in turn lead to more expensive industrial processes and subsequent increases in operating costs.

Figures on different measures such as energy consumed, percentages of recycled materials used, output of pollutants, etc. are frequently used to indicate the environmental part of the issue. But what about the business part?

From this general background it can be concluded that environmental considerations will play an increasingly important role in corporate behavior in the future. Thus, it is in the interest of management to integrate environmental and general performance information of the enterprise with each other. Technically, such integration can be made in three broad ways:

(a) Reductionist approach. Information available in various physical measures, financial figures, technical estimates, etc., are reduced to one of those measurement units -- typically money, energy or pollutants units;

(b) Holistic approach. All information is put into a single image -- typically an abstract numerical index; or,

(c) Structured approach. A regular use of a rigorous format (typically an accounting format) which facilitates the interpretation of various performance indicators and their use for decision making purposes.

Behind such technical differences more fundamental differences in the way to perceive a corporation may be found. Reductionist approaches may be worth consideration in compliance-oriented companies when sustainable development and environmental relations appear as marginal changes to ordinary business habits. Structured approaches, on the other hand, open up a whole field of radical possibilities, for instance to analyze the company as an ecological entity. If enterprises were to be analyzed as ecological entities, entirely new ideas would follow. New business opportunities might be identified as well as industry concepts, social dynamics and ways to understand economic relations.

III. CORPORATE ENVIRONMENTAL REPORTING

The study, *Environmental Management in Transnational Corporations* (United Nations, 1993, Sales No. E.94.II.A.2), reports on an extensive survey of transnational corporations. The report concludes that the national legal and cultural environment from which a transnational company comes from is significantly important for management style and enterprise strategy. Any place in the world is specific in the respect that "all business is local". To set the conditions for environmental reporting, and especially for discussing options for integrating environmental and financial information, much specific company information is needed. One has to understand the kind of environmental conditions which prevail to judge what comparisons can be made and what methods can be used.

In the literature there is not much empirically-based evidence on company practices involving the relationships between financial and environmental performance. The surveys conducted by the United Nations, for the most part, documented a low profile regarding any environmental disclosures among the transnational companies studied. The 1992 report, *Environmental Disclosures: Internationaol Survey of Corporate Reporting Practices*, (E/C.10/AC.3/1992/3) covers 222 transnational companies on accounting for sustainable development and environmental disclosure in general. Only 7 companies (3%) had amounts of their environmental debts in their annual reports. Other figures were more common:

(a) 14% had footnotes in their balance sheets on environmental conditions;

(b) 62% disclosed information on environmental improvements;

(c) 64% had text on financial outcomes of environmental problems in their annual reports;

(d) 70% told the general public about their environmental policies; and,

(e) 86% had at least something to say about the environment.

The United Nations report concluded that the general quality of public environmental information is low, never audited and very seldom quantified. There is no information for making comparisons between companies intelligible:

"...while transnational companies are aware of environmental issues, their disclosures remain qualitative, descriptive, partial and difficult to compare. Not only was there little quantitative information but often no time period was attached to the qualitative information. Moreover, there was no relationship drawn among amounts spent, results achieved and targets set. Therefore, it was not possible to gauge the environmental performance of the corporations, let alone the impact of their environmental activities on their financial results."[1]

The fact that the biggest companies do so little to raise the visibility of their corporate culture could be attributed to their low aspirations in the field, but, says the United Nations report, one should not underestimate the difficulties at hand. When estimating environmental debts and environmental costs one has to consider the following:

(a) The debts are uncertain because they are determined by future laws and regulations;

(b) They extend over a longer period than most debts and costs;

(c) Invisible critical thresholds make cost and debt functions discontinuous;

(d) Environmental damage is harmful to many. Management can never know absolutely which persons are affected, or how many of them may claim compensation;

(e) If an enterprise goes beyond the levels required by the law in its long-term responsibilities there will usually be a trade-off position against short-term profitability.

The United Nations study also points to differences in legislation among countries. Accounting practices differ from one country to another regarding what is accepted as ordinary costs, what should be registered as an investment and what is acceptable as a cost during an accounting year. Some countries allow extraordinarily short periods for depreciating or amortizing environmental investments for income tax purposes. There are also legal clauses, for example in Canada and Sweden, that force companies to accumulate provisions during the lives of assets for clean-up costs that will come at the end of the lives of those assets.

In the previously mentioned United Nations survey it was also reported that about one-third of the companies claim that they have environmental accounting of some sort, "...however, no examples of concrete methodologies were submitted."[2]

At the same time a more optimistic outlook about the issue is emerging. A recent survey by TRG Revision, a Swedish subsidiary of the accounting firm of Deloitte Touche Tohmatsu International, reported a considerably growing interest in environmental reporting among the companies listed on the Stockholm Stock Exchange, although no connections with overall company performance measures was discussed. In 1991 McKinsey & Co., an international consultancy firm, reported that out of 400 companies 92% perceived the environment to be one of the big challenges of the next century.[3]

Current developments in the "Total Quality Management" field may point to new possibilities (Bergström and Gummesson 1994; Welford and Gouldson 1993). The Eco Management and Audit Scheme (EMAS) recently promulgated by the European Commission is one such possibility. As the scheme is recent, it's practical implications are speculative: EMAS should raise environmental awareness in many companies as a first step. The programme is designed to be compatible with practices in the dominating industrial enterprises in the European Union. This means that it may counteract ambitions and functions in enterprises that are environmentally more radical.

When high environmental standards are a customer claim, the issues more easily become a part of "normal business". Public bodies such as government agencies, municipal administration, schools, etc. may serve as quality conscious forerunners on the demand side of the market. Standards set on office equipment, building materials, copy paper, etc. will soon expand to other demand sectors. The evidence from Sweden is strong on this point.

The research in recent years by the Stockholm House of Sustainable Economy and the environmental research group at Stockholm University School of Business has shown that site consciousness is one of the crucial conditions for a company to become seriously "green". Consequently radical practices in reforming economic concepts, accounting and reporting, etc., are first developed by municipalities, local housing companies and similar organizations with no alternative to staying with "their" environment. This study uses parallels to that research for discussing possible future developments in transnational corporations.

IV. CASE STUDIES

For the purpose of this report, six case studies were undertaken to provide a basis for forming conclusions as to the feasibility of linking environmental performance indicators with the financial and other information normally supplied by enterprises. The case studies were selected to represent different kinds of environmental problems encountered by enterprises and, of course, do not cover all possible cases. However, it is felt that these cases provide good potential for being able to formulate recommend ways to integrate business and environmental performance reporting.

All the cases follow the same pattern:

(a) A short general background is provided on the company and its position in the company Group. All of the case studies describe a Swedish unit as part of a transnational Group. Two are international Group headquarters (Volvo and Astra), and four are subsidiaries in a Group (Sunwing, WMI Sellbergs, McDonald's and Stora) of which two of the latter four have a Swedish parent company (Sunwing and Stora). In combination with some notes on company history this background provides the overall conditions for performance reporting by the company.

(b) Some information is given on company policies, strategies, operational targets etc., which provided relevant performance criteria for analysis purposes.

(c) There is a brief presentation on the industry which is necessary to understand the kinds of environmental issues that are appropriate to be raised. The cases are substantially different in this respect:

(i) WMI Sellbergs' environmental concern is central to the business idea of the company. Re-use and recycling is what is offered to their customers.

(ii) Sunwing Hotels manage hotel establishments around the world. The hotel industry has generally demonstrated awareness of the importance of good environmental management.

(iii) Svenska McDonald's is the Swedish branch of world's most well-known fast foods chain. The Group has made a well-recognized upgrading of it's environmental profile in many countries.

(iv) AB Volvo as a car manufacturer displays most of the typical features of the old industries. The company is known to employ environmental considerations in product design. Within the automobile industry, environmental arguments have become increasingly important in marketing products.

(v) STORA Skog is one of the largest forestry companies in Sweden within which a serious environmental debate is going on.

(vi) Astra is a very competitive company in medical drugs. The pharmaceutical industry operates under extreme testing, documentation and safety conditions. Companies are generally law-driven and conservative.

The integration of environmental performance indicators
with financial information by transnational corporations

47

(d) The resource relations of which the company is a part are documented.

The specific environmental issues which may be raised relate to the circulation of resources. Thus, the case studies report on the resources base, throughput of resources, and kinds of resources

transformations that are made by the company in accordance with the scheme outlined in figure II.1 above.

(e) Specific reporting practices in each company are reported as well as comments from management on the need for information and expectations for changes in the future.

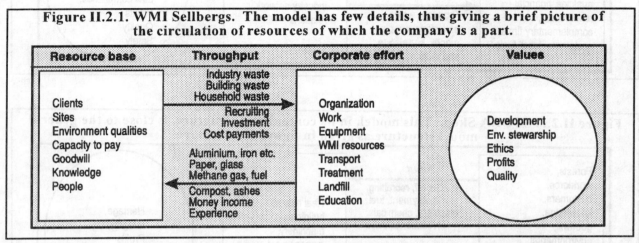

Figure II.2.1. WMI Sellbergs. The model has few details, thus giving a brief picture of the circulation of resources of which the company is a part.

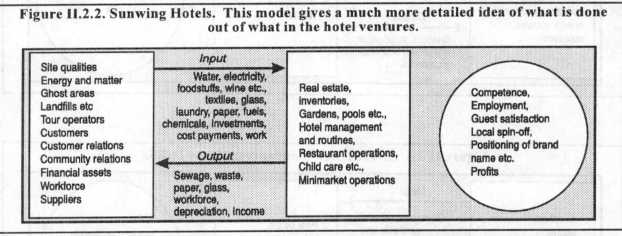

Figure II.2.2. Sunwing Hotels. This model gives a much more detailed idea of what is done out of what in the hotel ventures.

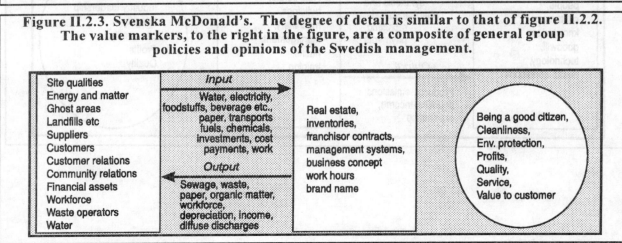

Figure II.2.3. Svenska McDonald's. The degree of detail is similar to that of figure II.2.2. The value markers, to the right in the figure, are a composite of general group policies and opinions of the Swedish management.

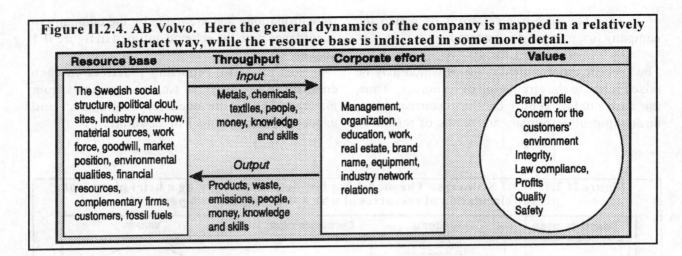

Figure II.2.4. AB Volvo. Here the general dynamics of the company is mapped in a relatively abstract way, while the resource base is indicated in some more detail.

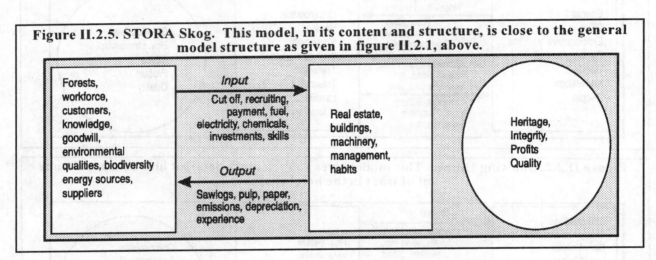

Figure II.2.5. STORA Skog. This model, in its content and structure, is close to the general model structure as given in figure II.2.1, above.

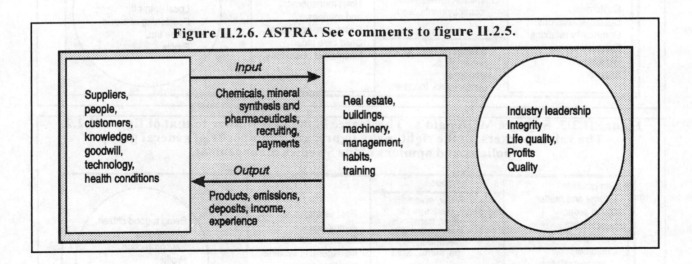

Figure II.2.6. ASTRA. See comments to figure II.2.5.

Using the brief model format as introduced in figure II.1, the six cases can be outlined as in figure II.2 (in 6 parts, 2.1 to 2.6) below. The specific structure of the six companies as economizing and circulating systems is described. Values in the right hand side of the figures are listed in alphabetical order.

V. REPORT DISCLOSURES FOUND IN THE CASE STUDIES

This section is a brief overview of current reporting practices in the case study companies.

The text is not intended as an assessment of the existing practices. It should rather be read as a factual record. It is not, however, a standardized record for each of the corporations studied but rather it illustrates where good information is available.

As indicated above, if these companies are integrated into transnational groups, they are exposed to the financial markets and generally have greater flexibility with regard to operations sites, and consequently are not expected to develop

Table II.1. A brief overview of reporting practices in the case studies

Question	Response
Are environmental issues a general management responsibility (distributed down the line)?	Yes, the issues are always acknowledged as such. With varying degrees of explicitness, Group management expects subsidiary units to take proper initiatives for the environment. Reporting and quality assurance are, however, handled by specialized units of staff.
Are environmental issues reported to the Chief Executive Officer and the Board on a regular basis?	Not always on a regular basis and not always identified as environmentally relevant. A fair answer is "most often".
Are there any safeguards against incorrect internal thinking about environmental conditions?	It does happen, but it does not seem to be the normal case. External auditors, external dissemination of information and certification schemes are in use or at least intensively discussed.
Are environmental conditions disclosed in information that is provided to the general public?	Only at a low level of information quality: policies and cost estimations occur alongside physical variables. But there is no environmental performance reporting.
Are environmental conditions integrated into general performance reporting?	No, but within some of the companies there is currently a discussion going on over the issue.

innovative practices regarding environmental reporting. Table II.1 below summarizes the environmental reporting information that was identified in the six case studies. Additional descriptions of the information are included in the documentation of the individual case studies.

A. The environment and general management

At Waste Management International Sellbergs AB (abbreviated as WMIS in this report) all line managers and certain designated individuals have full responsibility for environmental as well

as other matters within limits set by the line organization. The line managers also operate a computerized environmental compliance assurance system. A special national staff unit is responsible for interpreting government regulations and relevant scientific knowledge. Educational programmes are developed by the personnel department.

At ASTRA, key figures concerning data on significant environmental impacts are reported on a regular basis at each unit. These figures monitor the levels stated in government concessions and regulations as well as in internal goals set up at each specific unit. The frequency of reporting depends on the level of importance of the data from an environmental impact point of view, as well as the ability of the data to be measured in an accurate manner. At one site five to six different figures are being monitored on a daily basis with regards to both usage and discharge of heavy metals, solvents and chlorine.

Within the McDonald's Group, all performance reporting follows a worldwide standard. Reports cover traditional financial statements such as the profit or loss from operations, gross profit and return on investment, and some other performance indicators such as the number of guests, sales per guest, cost of food as a percentage of total costs, etc. This information is supplemented with figures on deliveries of bread-buns, french fries, etc. A goal of 90 per cent waste reduction is measured once every year by an inventory procedure. No external environmental auditor is engaged.

Regarding the qualitative goals which restaurants in the world aspire, there is no reporting from Sweden to McDonald's Corporation by traditional means. In the future, however, a number of key ratios will be developed and reported in order to demonstrate compliance with the company's environmental goals. This year, a key ratio regarding the usage of polyethylene plastic will be put in use. The ratio will make it possible to understand and direct the flows of recycled materials. Concerning distribution, besides actual costs, kilometres per ton distributed (outgoing transportation) is used as a key ratio.

Each Sunwing hotel has a number of standard financial and operational ratios that are used as guidelines. Such ratios are: the number of occupied beds and rooms, average number of employees, and total revenues per guest-night. Every week each hotel reports their liquidity (cash balance) and results of operations before taxes to the head office in Stockholm. On a monthly basis a profit or loss statement, a balance sheet, a statement of cash flows and the above-mentioned financial and operational ratios are also reported. All reports are entered on forms and sent to Stockholm by mail or by facsimile. Sunwing hotels further report to Group headquarters with the same frequency and the same content. No information goes the opposite way,

i.e. from Group headquarters (Airtours) to the head office of Sunwing hotels and back to the individual hotels. However, feed-back is given when divergence from planned budgets occurs.

It is customary in the hotel business to record a ratio of the quantity of water and electricity consumed (with the number of guests as the denominator) and this has been done for many years in order to keep costs at a low level. This is, however, not regarded as environmental reporting and it is not reported to the Stockholm head office.

Volvo has developed a life-cycle analysis (LCA) system called EPS (Environmental Priority Strategies in product design). EPS helps the Volvo companies to examine the environmental impact of materials and products. Volvo also has an environmental data base, MOTIV, which contains information on more than 5,000 chemical products. A central group in Volvo, called Yrkeshygien (Work environment hygiene), supports the system and evaluates new chemicals. The data base can, for example, be used to find all products that contain a certain chemical. This can be of help in an emergency situation.

Volvo's current environmental policy pledges:

(a) to develop and market products with superior environmental properties and which meet the highest efficiency requirements;
(b) to use manufacturing processes that have the least possible impact on the environment;
(c) to actively participate in and conduct research and development activities in the environmental field;
(d) to select environmentally compatible and recyclable material in connection with the development and manufacturing of their products, and when they purchase components from their suppliers;
(e) to apply a total view regarding the adverse impact of products on the environment; and
(f) to strive to attain a uniform, worldwide environmental standard for processes and products.

Every company manager is responsible for implementing actions consistent with the above policies. The policy in itself is, however, a centralized product. Volvo's Environmental Council has the final responsibility for the policies. The Council is comprised of specialists from the Group's head office and from the various operating units, and the organization is responsible for the ongoing coordination of activities. In a new version of the policies, suppliers doing business with Volvo must also comply with the above environmental requirements.

B. Environmental audits and internal environment reports

In WMIS the national environmental audit unit is separate from the line units. The audit unit reports directly to WMIS Group headquarters. Special measures safeguard the integrity of the unit from all of the other operating units. The National Environmental Audit Unit maintains a compliance verification system, a database covering all contracts, rules, and claims, etc., which may affect what is the proper thing to do in specific situations. The audit unit assess managerial and auditing procedures used by the units and sets internationally valid standards.

The environmental auditing at ASTRA is primarily conducted by the company's own personnel visiting the different units. However at ASTRA's units in Södertälje, auditing is done by an external auditor. Internal environmental reports are presented to the Board of Directors whenever any changes in production takes place -- otherwise the Board receives no formal environmental reports on a regular basis other than the corporation's annual reports.

The environmental auditors at Volvo are in-house specialists supervised by the Group's environmental auditor. Auditors need to have access to all accounting and other records with nothing hidden, which is the main reason why external consultants are not used. Still, the auditors always come from units outside of the unit being reviewed. Only on rare occasions have the auditors been from outside Volvo, and this happens solely when special legal knowledge is needed.

The audit focuses on different levels of environmental issues:

(a) Compliance with current national legislation;

(b) Compliance with probable future legislation; and

(c) How the Volvo environmental policy objectives have been translated into action plans and concrete measures in the company.

The environmental auditors do not perform an annual audit. They rely on a surprise audit to get a maximum reading on policy compliance. The audit is more of an evaluation than a traditional audit in the financial sense. The environmental audit does not give any answers as to how the company should improve its environmental programme: the auditors only point out where poor environmental performance was found. It is the appropriate company manager's responsibility to take suitable measures.

C. Disclosure of environmental information, external certification and other matters

A discussion of external certification and environmental auditing is on the agenda at STORA Skog. The Chief Executive Officer has stressed the importance of enjoying public confidence in how forest industries are managed in general, and especially from an environmental point of view. Today STORA Skog has no external environmental auditing. Third party certification is not so much about just attesting to the final product as it is to get a neutral third party's view of the complete operating cycle, from the trees in the forest to the final output.

An environmental audit report was included in the 1993 ASTRA annual report. Starting in 1994 it will be presented as a separate annual report.

WMX Technologies Group, the parent company of WMIS, publishes an annual environmental report. It covers all subsidiaries and is a well-written record which shows important changes in capacities, procedures and conditions of environmental relevance. Very little is said about actual performance except in the regulations compliance part, where the following statement appears:

(a) Percentage of areas audited where no significant compliance issues were found;

(b) Percentage of issues resolved according to schedule;

(c) Percentage of penalties which were from self-reported incidents; and,

(d) A compliance index which relates federal penalties to gross revenues.

An independent consultant, Arthur D. Little Inc., performs an external audit which covers policies and procedures, organization, training programmes, regulatory and management reporting systems, performance incentives and disciplinary action programmes, regulatory surveillance systems, audit programmes and corrective action systems, and other environmental management programmes and systems in place throughout the company and its subsidiaries.

D. Integration of environmental and financial performance measures

There is no reference to traditional business performance measures in the annual environmental report of WMIS (WMX Technologies). Environmental concern is treated as an important corporate item although separate from general business considerations. The management at WMIS emphasizes that integration of financial and environmental measures will be needed in the management processes, but they lack a valid way to construct such measures.

At ASTRA key figures of environmental significance are not related to financial factors -- i.e. there is no integrated environmental and financial reporting. The management does not foresee any such immediate integration even though they have access to a sophisticated data base.

The question of how to integrate financial and environmental reporting is differently stated at STORA Skog. This integration is a difficult and urgent question for a forestry company. The value of the forest stock is one of the important figures in the general financial report. This stock has a book value but the real value is to be found in the ecological conditions in the forests themselves. In the near future there will be an increasing need for standardized, reliable and confident methods for integrating environmental information with established financial information.

McDonald's is, at least for the Swedish operations, approaching a "cyclical society policy". This implies that all operations should be made in accordance with the following four basic principles:

(a) Minimal usage of underground mineral deposits (including fossil   fuels);

(b) Artificial compounds should not be used;

(c) The physical conditions of the eco-system must be preserved; and

(d) The energy usage (metabolism) of society must be reduced equitably.

There is obviously some way to go before these requirements are fulfilled. Svenska McDonald's in the future will probably include three kinds of capital in their performance reports, which they preliminarily call "financial, natural and human capital", in order to keep track of the direction which they have decided to follow.

VI. THE POTENTIAL FOR INTEGRATING BUSINESS AND ENVIRONMENTAL PERFORMANCE INFORMATION

A. Potential for developing environmental reports in monetary terms

As indicated earlier in the report there are various ways to integrate notions of environmental performance into general business performance concepts. Several studies focusing on general industry statistics and similar topics employ add-on strategies where traditional financial accounting is given. However, no practical managerial applications of such a methodology were found in the literature when conducting the research for this report. It might seem that there are limited prospects for developing environmental reports for managerial use in monetary terms. Still, the approach is in high esteem in academic and political institutions.

The integration of environmental performance indicators
with financial information by transnational corporations

53

B. Potential for developing holistic performance indexes

Another alternative is Life Cycle Analysis. This technique maps all environmental relationships of a product or a process when all of its components are traced back to their natural beginnings, called the "cradle", and all applications are traced forward to their ultimate dispositions, called their "graves". It is an engineering approach in which the aim is to minimize harmful environmental outcomes. The user of LCA has to face delicate valuational decisions since various effects must be combined. How, for example, do you evaluate shorter transport distances against greater sewage volumes?

A radical development of the technique has been done for the Swedish car industry, called the EPS system (Ryding and Steen 1991). This is described in the case for Volvo. By means of standard procedures, an "environmental index" can be computed for every part in a construction (or production) process. The indexes of all the parts are then multiplied to produce a compound index for the whole process. To standardize the results, all health outcomes are related to a single individual and all ecological outcomes are related to a single square kilometer.

The EPS approach is rough and it has been criticized for its superficial way of handling value questions. Its merit is that it produces a specific number, which can work as a benchmark for further and more specific investigations of environmental outcomes. Probably a lot of environmental improvements never come about because there is no specific place on where to start. If such an abstract index as EPS measures should be built into managerial control systems, the opposite effect is likely to occur. Managers will not have ways to go beyond the numbers -- only experts can tell what questions to ask. The importance of performance indicators can be concrete and intelligible to those using them. According to Johnson & Kaplan (1987) this is exactly the reason behind the current dominance of financial indicators in corporate decision making.

Index methods are sparse outside engineering practices. One exception may be rating models as those used by, for example, Standard & Poor's and Moody's. Standard & Poor's describe their model in this way:

"Typically, analyses are based on five years of historical results and projected data ranging from two to three years. Projections are important for discussing the firm's planning process, future direction, and management philosophy. They are not intended to be used to judge management's forecasting ability.

Ratings compare risks among debt issuers, with relative ranking taking place on several levels. Issuers in the same industry are compared on both a domestic and international basis. S&P's analytical approach is designed to produce a cross-border yardstick of comparison.

Specific weights are not assigned to any of the rating criteria. The rating finally assigned is a synthesis of qualitative and quantitative factors discussed within a committee on a case-by-case basis.

S&P's analytical framework for corporate issuers is comprised of two key components. The first is oriented towards business or competitive analysis; the second is related to financial analysis. ..." (Standard & Poor's, 1992)

In the presentation of their rating criteria Standard & Poor's underlines that the rating process is not limited to an examination of various financial measures. On the contrary, a proper assessment of debt protection levels requires a broader perspective involving business fundamentals. They do not explicitly point to environmental conditions as being part of those business fundamentals, although nothing is said about keeping such considerations out.

It is likely that the environmental conditions for a corporation are growing in importance when conducting a rating. The very concealed nature of the process where only the outcome, the rating, is disclosed leads to speculation. Rating companies claiming they make explicit environmentally-based judgments have appeared on the market recently. This may force a more general recognition of environmental concern as being an integral part of business management.

C. Potential for refining the structure of performance information

Among the reported case studies, this strategy is discussed at least within two of the case study companies, Svenska McDonald's and STORA Skog. In the future, Svenska McDonald's will probably include "financial, natural and human capital" in their performance reports. To do that in a practical way they will have to adopt an economic model of the company where the capital concept is open for this kind of extended use. The basic scheme used in the case studies illustrates exactly that: since the company is circulating all kinds of resources (financial, natural/material, and human/social) and acts as an economizing unit, it seems to be a good thing not to confuse the various kinds of resources with each other.

The experiences from research and practical developments in Swedish municipalities underline the same thing: a site-conscious (and/or quality-conscious) management system is dependant upon economic concepts where all kinds of resources and all kinds of qualities are rigorously treated. Basically, being "economical" is to consider how means are used to meet ends. Therefore, management should consider how to use scarce resources for attaining explicit goals, which implies claims for effectiveness, claims for thrift, and claims for good margins and low risk financially as well as for outside aspects of ventures.

In Sweden a new kind of accounting system has been developed and tested that intends to be directly derived from sustainable development visions and strategies. Performance measurement is tied to the concepts presented above: effectiveness, thrift and margin. The approach is called SDR (sustainable development records). Various general conditions restricting the approach are presented elsewhere (Daly 1977, Bergström 1992, 1993, 1995). The general SDR model structure, which is the one used in the case studies for this report (see figure II.1), is mapped into an

accounting scheme where the stock data end up in a balance sheet and change-process data in a profit-and-loss account.

From an environmental point of view the SDR approach differs from most other approaches in not establishing separate environmental accounts. The idea is to develop measures for general strategic performance assessment in a way which facilitates concern about the environment. The problem addressed with SDR is not the problem of giving the environment due attention but to stop treating the environment as something peculiar, where "normal" ways of being rational do not apply.

SDR is thus an accounting approach to sustainable development. It is currently being tried out in several Swedish municipalities, state agencies and companies. The basis of SDR theory is that every operation is defined by its ecological conditions and these conditions are as intelligible as a business firm. From a business firm you expect financial results, cost controls and accountability. SDR is thus focused on how to organize information on essential results from enterprise efforts. Technically the method is built upon a rigorous implementation of double entry accounting in real terms, i.e. the entries do not have to be made in monetary measures.

SDR theory is built on a combination of three sets of logic with roots in the fields of ecology, economy and accounting.

(a) The ecological logic or the scheme of accounts and other descriptive concepts are built on the physical transformation process of the business;

(b) The economical logic whereby:

(i)   Explicit value criteria are added to the circulation model. This is the foundation for tracking revenue concepts.

(ii)   The essence of economic analysis is a non-violence principle, meaning that one should never use more effort than needed to reach a specific target.

(c) The accounting logic whereby:

(i)   Double entry accounting means that every transaction is done both at the source and at the destination. In that way detailed records are kept of the throughput generated by the business.

(ii) Balance sheets (there is one for every kind of measure used) and income statements show a clear distinction between asset levels and positional measures on the one hand and throughput and change measures on the other.

(iii) When the books are closed debits and credits must equal, which makes it possible to audit the system systematically. Because of this one would be willing to handle very complex information which otherwise would break the system down.

The SDR approach is focused on managerial accounting and control systems. Experience so far shows some benefits with the system: first, the system development process has worked as a vehicle for making strategic thinking more explicit; second, the combination of real measures and explicit performance criteria gives room for quantitative following up where only personal subjective judgment was articulated before. Important qualities (technical qualities, environmental qualities, customer qualities, etc.) are handled in the same way, and with the same rigor as financial data. Third, sub-parts of a business as well as development over time are made comparable. Fourth, when discussing sustainable development the SDR approach can measure "small steps in the right direction" even if those steps are taken at a very low level. Results measures in the SDR system are mostly of a key indicator type, as illustrated in Table II.2. All key indicators are designed to give a higher number for a better state of the system.

Table II.2. Possible key indicators of the SDR type. The examples refer to the case studies and correspond to the models in figure II.2, above. It should be noted that these key indicators are presented here for illustrative reasons only. They are not sanctioned by the management of the enterprises in the case studies.

	Effectiveness = A value indicator/a corporate effort indicator	Thrift = A corporate effort indicator/a throughput indicator	Margin = A throughput indicator/a resource base indicator [1]
(WMI Sellbergs)	Profits/WMI Resources employed	Salable matter/ total throughput of matter	Liquid assets/cost turnaround
(Sunwing hotels)	Customer satisfaction/ real estate value	Number of guests/laundry	Local workforce/employees
(Svenska McDonald's)	Profits/capital employed	Capital/transport volume	Recycled waste/material input
(Volvo)	Brand position/cars sold	Cars produced/emissions from factories	Critical emission/current emission
(STORA Skog)	Market share in highest price segment	Sales/transport volumes	Organic growth/cutoff
(ASTRA)	Market share for the main product line	Sales/marketing effort	Patents/distributed profits

VII. CONCLUSIONS

A. Differences between industries

Some of the arguments around the general conditions for an enterprise to develop practical environmental awareness are summarized in table II.3 below, where the "state of advantage" column summarizes those conditions which would make advanced and ambitious approaches to the integration of business and environmental performance measures likely.

Table II.3. Conditions for an enterprise to develop a practical environmental awareness

Type of condition	State of advantage	State of disadvantage
Site dependency	Clear and understood	Confused or unknown
Exposure to financial markets	Low	High
Formal integrity	High	Low
Abstract vs. material operations	Material	Abstract
Customer claims	Quality conscious	Diffuse, price conscious

From these classifications it can be concluded that the cases chosen for this report exist in a middle ground between the advantage and the disadvantage states for developing new practices around financial and environmental reporting.

B. Importance of market and legislative contexts

Environmental costs to enterprises are a function of time to adapt. Tough political measures in combination with early warnings before the needed changes actually occur seem to be a successful political strategy for influencing companies with mainly reactive and passive ways to handle environmental and developmental issues.

It appears that transnational corporations typically need firm legislation before they will develop sustainable practices. This is partly because enterprise managements feel that the internalization of environmental costs in their costs of production will reduce the profitability of their products and make them uncompetitive in the market place and will have an adverse impact on the profitability of the enterprises. At the same time TNCs can usually avoid strong legislation because of their international flexibility in locating production facilities. However, this flexibility should not be over emphasized since investments are made somewhere in the world and economic considerations such as transporation costs and costs or relocation could significantly affect the profitability of operations.

State agencies and local municipalities should not under estimate their power as buying agents. Where a unified policy concerning environmental standards is possible, these bodies constitute a considerable market for many industries. An illustrative case was the development of low-chlorine/chlorine-free paper in Sweden.

It has been found that companies with an ambition to have a more sustainable development become increasingly involved with non-governmental organizations (NGOs). STORA Skog has good forestry management and social reasons to make their production more adapted to ecological sustainability. Partly, this can only be accomplished at a cost which their markets will accept only if environmental NGOs push both their competitors and their customers in the same direction. The legitimacy of NGOs as agents capable of raising "necessary" claims is thus an important asset for companies with advanced environmental concepts.

EMAS and similar initiatives will work both as a stimulus and as an impediment to a sustainable development. On the one hand such programmes initiate action where environmental initiatives otherwise probably would be delayed. On the other hand it is expected that conservative business practices and the slowness of change will be consolidated and institutionalized.

C. What may and may not be accomplished in the future

The point about EMAS mentioned above illustrates a general condition worth further mention. Some methods and practices may be successful in introducing the environment on to the business agenda but at the same time the actions taken may be obstacles for going beyond a first step. It takes confidence to initiate an action programme. Those trying to take a first step in raising environmental concern may find a given action most valuable, while the same action may be criticized by others as being an impediment to going further. The context within which environmental and financial performance is integrated differ considerably.

Items which need to be dealt with in the future may include the following:

(a) Cost/benefit analysis, full cost accounting and similar methods aimed at representing environmental relations in monetary terms rest on methodological assumptions which are still evolving;

(b) Environmental reports which are not related either to enterprise policies or performance open up the field and give some discretion to employees with interest in the issues. These iniatives, whenever appropriate, should be integrated with management responsibilities to capitalize on and encourage employee participation;

(c) Heavy dependence on experts in decision-making bodies for interpreting the information given is a consequence of abstract figures and long chains of derivation. Greater efforts are required to ensure that the information is presented to decision-makers in a user-friendly manner;

(d) Governmental bodies, international regulation and other policies provide many options for action. Table II.4 describes the lines of public action which would support activities of enterprises whose management styles have different characteristics and goals;

(e) When sustainable development and environmental considerations become the central concern of management it is possible that values which earlier were taken as given may need to be reassessed; and,

(f) On the one hand some enterprises may introduce the kind of "revolutionary" processes which will follow their acknowledgment of broad environmental issues. Alternatively, other enterprises may be tempted to use inexpensive investigations, a low profile and a compliance strategy while others make the investments. The most suitable approach has yet to be defined.

Among the progressive routes, the following are emphasized:

(a) Authorities and the research community can help enterprises that exhibit motivation on their own to develop new ways to integrate various kinds of performance information. Standards for external disclosure will be successful if they build upon principles and techniques which have worked well in internal management systems;

(b) Generally, it seems that structural approaches, especially the broadening of basic economic concepts, are more open-ended and thus innovative than the alternatives. Such strategies should not, however, be expected to be quickly adopted in transnational corporations generally. Those with well-recognized site and resource problems to handle are expected to set the pace;

(c) Generally, government policies would benefit from a more flexible attitude to management realities in TNCs. Table II.4 summarizes observations and conclusions made in this report showing a wide assortment of possible public actions that would support the integration of environmental considerations into corporate performance concepts, whether performance measures are publicly disclosed or not. The general format of the scheme, as well as many of the specific points in the table, are taken from the Benchmark Corporate Environmental Survey.[4]

Table II.4. The relationship of types of enterprise managements to their operating activities and to governmental actions which support those activities

Management type	Enterprise activities	Supporting governmental activities
I. Compliance-oriented management *(The reactive enterprise)*	End-of pipe solutions Abatement procedures Compliance reports Environmental experts in staff unit Emergency response	Command and control, realistic regulation Dialogue with industry organizations, conferences etc. Inform on regulations, early attention Tough enforcement
II. Preventive management *(The lean and precautionary enterprise)*	Internal audits Pollution prevention Waste minimization Public information Green accounting Line management responsibility for environmental issues.	Increased liabilities Waste treatment requirements, restrictive landfills policies, etc. Community right-to-know claims Energy conservation programmes, demand side management Green tax schemes
III. Strategic environmental management *(The concept seeking enterprise)*	Public dialogue, environmental product positioning External audits and use of environmental certification programmes Disclosure of quantitative throughput information; extensive LCA programmes Integration of environment, health and safety into technical design, "green" R&D	Stable regulatory build-up Green labeling programmes Support of consumer and green investor initiatives Strategic buying projects within public bodies Build-up of local and regional bodies for focusing joint industrial - public throughput issues
IV. Sustainable development management *(The responsive enterprise)*	Acknowledging of the enterprise's role in international wealth distribution Acknowledging of the enterprise's role in throughput of matter and energy Policies for ethical sales, climate change, throughput thrift etc. Use of best practice in all operations internationally. Sustainable development fully integrated into enterprise performance reporting. International auditing.	International information dissemination International harmonization of environmental regulation, standards and taxation. Health, safety and sustainable development is prioritized before traditional free trade values in international negotiations.

ANNEXES

CASE STUDIES

Annex I

WMI SELLBERGS AB

WMI Sellbergs AB (WMIS) is a wholly owned subsidiary of Waste Management International plc, which in turn is a subsidiary of the United States based WMX Technologies Inc. group. WMIS was established in 1882. Under former owners WMIS was a well-established company in the transport, garbage collection and building sectors in Sweden. During the period 1965-1980 activities outside garbage collection were disposed of. When the company was taken over by WMI (in 1989) a new management and a new business concept direct the operations. In general terms the change was from a focus on garbage collection to a focus on environmental services.

Garbage collection and transportation away from residential areas was a logical kind of activity when society was thought of as a linear process where physical matter was removed from nature at one end, converted and sold as products, consumed or worn down and, finally, thrown away as waste at the other end. As the open and empty world increasingly appear to be a closed and full world, this attitude ceased to work practically. This is recognized everywhere in the industrial world these days and the concept of the eco-cycle society is thus generally high on the agenda. WMX/WMI management saw early the challenge in these changes and established the new service concept which is now implemented in all its subsidiaries.

WMIS adopted that concept in 1991. The company provides technical solutions for residual products management, environmental protection and enhancement for industry, the building sector and municipalities (households). The following terms are used as strategic guidelines:

- o   Environmental stewardship
- o   Ethics
- o   Quality

- o   Development
- o   Profitability.

The company is a recognized stake holder in the development of public environmental policy in Sweden. It is a drawback to the company that rules around waste dumps are quite liberal and that enforcement is weak. WMIS would like to see more stringent regulations forcing dumps either to close or to be engineered to higher standards. Such engineered facilities are most often referred to as landfills. Recycling will be a less (or non-) profitable business as long as the full costs of dumps are hidden. Regulations around incineration are stricter, which is to the benefit of the company. There are almost no regulations around recycling due to short history of the phenomenon in its modern urban context.

Compared to most other countries, Sweden is sparsely populated -- 19 persons per square kilometer. The difference between Sweden and other European countries in this respect is more than a factor of ten. This means that there is substantial space for dumps and long transport distances. Garbage volumes are at a typical European level in an international comparison: every Swede produces (on the average) 1 ton of waste each year to be transported somewhere.

The competitive situation in Sweden is somewhat complex. Because of the changes in business concepts both within WMIS and within other companies it is not known who is competing with whom and about what. If we restrict the discussion to the traditional industry definition (garbage collection), WMIS is the biggest company in its field in Sweden. The bigger municipalities handle their own garbage and the local regulations in the sector differ substantially.

Assets and operations

The company owns facilities for recycling and treatment of all kinds of materials. A few landfills are owned by the company. In total the company has personnel at 40-50 sites. The turnover is close to a billion SEK and the profit margin on operations is between 10 and 15 %. The operations depend on material assets worth 0.6 billion SEK and around 1150 employees. The figures show that most work is of a manual service type.

The parent company (WMI) provides research and testing facilities as well as management systems suitable for implementing the business concept.

Throughput

Table II.A.I.1. Volumes of material throughput in Sweden, as various forms of residual products. Figures from 1990 with a forecast for 2010.

(millions of tons)	The year 1990	The year 2010
Recycling of matter	1,5	3,5 (200-300 facilities)
Energy extraction	1,7	3,5 (40-50 facilities)
Landfill/dump	6,8 (450 facilities)	3,0 (20-30 facilities)
Total amount	10	10

(Source: WMIS)

The national figures for materials throughput for 1990 with a 20 years forecast are shown in Table II.A.I.1. In Table II.A.I.2 the current input-output pattern from a WMIS point of view is shown. As can be seen the linear pattern is still there. Differences among local regulations impede rational procedures for facilitating reuse etc. in large scale. Where local/regional integration has been possible, a higher degree of recycling does occur. The BRINI technology, which is implemented in suburbs east of Stockholm, leaves only 10% of the input to landfill. Figure II.A.I.1 shows that the technology compensates for a low degree of source separation with a high degree of treatment. A higher degree of

cost effective recycling in the future depends heavily on changes in household habits.

Table II.A.I.2. The national input-output structure of residual products in Sweden. 1991 figures. All numbers are millions of tons.

Destiny or output:	Recycling of matter	Energy extraction	Landfill/ dump	Total input
Input sources				
Households	0,48	1,44	1,28	3,20
Industry	1,00	0,25	3,75	5,00
Building	-	-	1,80	1,80
Total output	1,48	1,69	6,83	10,00

(Source WMIS)

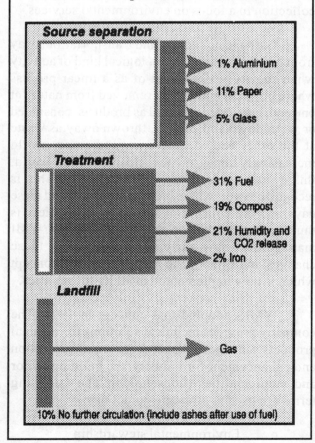

Figure II.A.I.1. An example of large scale technique for recycling. The figures are from WMIS' Kovik plant east of Stockholm.

(Source: WMIS)

Manpower throughput is indicated by the number of employees (see above). A special kind of resource is knowledge, which has all facets in common with other resources: it has a source, it needs support, maintenance and regeneration and there needs to be a certain capacity to make use of it. During the last year WMIS has invested in a management training programme for all managers focusing on required changes in leadership and skills when moving from waste handling to environmental services. This programme was followed by a basic ecological education programme for all employees, which is a way to build up capacity and thus flexibility to react in accordance with company policy.

Financial data is used to determine operating costs, amortization and investments. Half of the non-operating payments are generated by surplus on business operations and half is attracted from external sources.

Figure II.A.I.2 gives a general model of the company as an economizing unit.

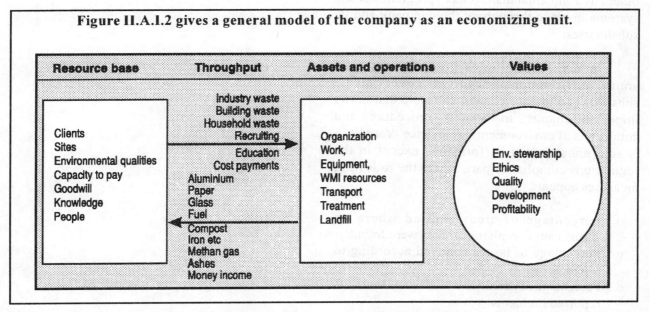

Figure II.A.I.2. The general structure of WMI Sellbergs AB. The figure focuses on resource circulation: Throughput is the flow of resources directly or indirectly managed by the company. The Resource Base is the environment of the company which has to exist to make the Assets and Operations valuable.

Report routines etc.

The WMI management system distinguishes 5 levels of responsibility for environmental performance:

1) All *line managers and individuals* have full responsibility for environmental matters as well as other matters within limits set by the line organization. The line managers operate a computerized environmental compliance assurance system.

2) A special national *staff unit* is responsible for interpretation of regulations and relevant scientific knowledge. Educational programmes are developed by the personnel department.

3) The *national environmental audit* unit is organized separately from the line units. The unit reports directly to WMI group head quarters. Special measures safeguard the integrity of the audit unit in relation to the rest of WMIS. The national environmental audit unit maintains a compliance verification system, a database covering all contracts, rules, claims etc., which may affect what is the correct thing to do in specific situations.

4)   The *international environmental audit* unit assess managerial and auditing procedures used and sets internationally valid standards.

5)   An independent consultant, Arthur D. Little, Inc., performs an *external audit* covering policies and procedures, organization, training programmes, regulatory and management reporting systems, performance incentives and disciplinary action programmes, regulatory surveillance systems, audit programmes and corrective action systems, and other environmental management programmes and systems in place throughout the company and its subsidiaries.

WMX Technologies group prepares an annual environmental report. It covers all its subsidiaries and is a well-written record that shows important changes in capacity, procedures and conditions of environmental relevance. Very little is said about actual performance except in the regulations compliance part, where the following measures appear:

o   Percentage of areas audited where no significant compliance issues were found.
o   Percentage of issues resolved according to schedule.
o   Percentage of penalties which were from self-reported incidents.
o   The IRRC compliance index, which divides federal penalties by gross revenues.

All these measures relate to conditions that exist in the United States. Also referring to the US scene, there are figures on worker accidents and injury rates in the Risk Reduction part of the report. There is no reference to traditional business performance measures in the annual environmental report. Environmental concern is treated as an important corporate item though separated from general business considerations.

In an interview with the Chief Executive Officer of WMIS, Mr. Bo Antoni, it was emphasized that integration of financial and environmental measures is needed in the management processes, but still there is no known way to construct such measures. In addition, he hesitates to publicly disclose such figures, since

"they must be relevant also to the reader with no training in the field. We cannot invite people to misinterpret us", says Mr. Antoni. "In this respect traditional financial figures have the advantage of being fully defined by human convention and to have a long history."

Annex II

SUNWING HOTELS

Sunwing Hotels AB is a wholly owned subsidiary of Scandinavian Leisure Group (SLG). Until April 1994, the Group was owned by the national airline of Sweden, Norway and Denmark; SAS (Scandinavian Airline System). As part of SAS's strategy to concentrate on its core business, SLG was sold to the British travel company Airtours Plc. Together, Airtours and SLG constitute the world's third largest leisure travel organization, with more than 3.5 million customers a year.

When the charter business (package tours, primarily to warmer countries in the Mediterranean) was introduced in the 1960s, the Scandinavian market was among the earliest to expand. Being among the pioneers, the Scandinavian Leisure Group has held a leading position ever since. Today, Scandinavian Leisure Group primarily consists of three parts that depend on each other and supply each other with business and customers:

- The tour operations - marketing, sales and arrangement of package holiday tours. In 1993, the tour operators of Scandinavian Leisure Group (Ving, Always, Saga) carried over 700.000 guests. Every third Swede or Norwegian who buys a holiday package chooses Ving for an annual vacation.
- A charter airline (Premiair) - flying the customers to the destinations. In 1993, the number of passengers carried was close to 1.4 million. The airline is jointly owned with a Danish tour operator, the Spies Group.
- The Sunwing Hotels - responsible for the SLG's owned and managed hotels. In 1993, the number of guests at the Sunwing hotels was 190,000 - of whom 140,000 were Ving customers (and the remainder from tour-operators that do not belong to the SLG group).

General context

The Scandinavian charter market is known for severe competition due to over capacity. Prices often have to be kept low which, in turn, erodes the business operating margins. The customer demand is, over the longer term, very dependent on the general economic situation and, over the shorter term, such things as present weather in Scandinavia (if the summer is rainy, the demand goes up and vice versa). When the demand falls, the tour operators have to lower their prices while at the same time being tied up with contracts for hotels and flight seats. The depreciation of the Swedish krona in 1992 has made the cost situation even worse.

The circumstances have forced, over the last few years, several Scandinavian charter airlines and tour operators into bankruptcy. For SLG's part, 1993 was the first year of a loss from operations in the company's history.

Strategy

In the 70's, SLG started to take over the operations of some hotels -- and later bought some of them -- in order to be able to influence and direct the occupancy; thereby being able to offer a higher degree of service to their own customers, as well as achieving a higher degree of control. Along with this reasoning, arguments such as cost reduction gains from co-ordination and higher value added services also can be pointed out.

The above situation accounts for the whole Group and Sunwing's strategy can not be easily separated. However, if anything should be pointed out for Sunwing alone it would be what Sunwing calls its concentration on "the Latin family" -- the whole family, including the children and grandparents. Staying at one of the Sunwing hotels should be convenient for the whole family. Practically speaking, the kind of guests who have chosen to stay at one of the Sunwing Hotels in the first place expect certain standards. As an example, the customers expect the same requirements surrounding children's safety as would have been found back home in Sweden or Norway. The guests also expect old newspapers to be collected and recycled as well as boxes for old batteries, etc. (The latter expectations make up one of the reasons why the environment came to be part of Sunwing's strategy -- see below.) Actually, the hotels are to be seen as whole "concepts" including such things as "the Miniclub" (day-care with Swedish staff), activities, excursions, shows and entertainment, food and beverage. An internal expression for these expectations trying to be met is: "Scandinavian content in Mediterranean wrapping-paper."

Regarding the new ownership, Airtours has stated that the operations of the Group will continue in their present form. The well-known Scandinavian brand-names will be kept and the clients are not supposed to notice any changes in image or service. The management will also remain unchanged.

Corporate effort

Sunwing operates 14 hotels in five countries: Cyprus, Gambia, Greece, Spain and Sweden. In total, eight of the hotels are owned by Sunwing, four are rented and the 14th (which is one of Sweden's best known golf and conference hotels) is operated on a management contract. In 1993, the number of employees were 1,468 and the turnover was SEK 597 million. The total number of beds was 8,028 at the end of 1993.

The operations depend on assets worth SEK 750 million. For 1993, the gross operating profit was 30 percent. That year, the bottom line showed a loss (after taxes paid) amounting to SEK 12 million.

Environmental protection

Sunwing started their environmental efforts in 1991. In the marketing-communication this is, however, not widely spoken of since the Scandinavian customers are regarded as being conscious as well as critical. According to the product manager of Sunwing, Miss Charlotte Wiebe, the environmental protection carried out is progressing continuously, but not until they have achieved higher standards will the hotels be marketed as "environmentally adapted".

Looking for reasons why environmental protection originally came to be included in the company's strategy, the following events can be pointed out: Firstly, in 1991, a customer of Sunwing, the German tour operator Neckerman, contacted Sunwing asking about environmentally-adapted hotels as they recognised that demand among their German customers. The German operator, which is the fourth biggest in the world, wanted to communicate the availability of green lodging in their brochures and -- more or less -- required a change. Secondly, in 1991, the International Federation of Tour Operators pointed out a number of areas in Spain where pollution -- if no preventive measures were taken -- would make tourism impossible within a period of ten years. The Spanish authorities took it seriously. Likewise -- although Sunwing didn't have any hotels situated in the areas in question -- Sunwing's hotel managers in Spain gave the problem a priority position in their agendas. Thirdly, as mentioned earlier, the Scandinavian customers are nowadays taking a certain degree of environmental services for granted.

As a fourth reason, a far-reaching environmental policy brought out in 1993 by Thomson, the world's biggest tour operator, can be mentioned. Thomson was also one of the first operators to hire an environmental manager and has now come up with detailed recommendations to all its suppliers and business customers. (Also TUI, the world's biggest tour operator, has a far-reaching environmental programme.) According to Miss Wiebe, Thomson's efforts serves as a model for the whole industry in this respect.

At the practical end, the environmental protection work started out with minor and diverse steps taken by a few of the hotel managers. The possibilities for what kind of action that can be taken differ between the countries in which Sunwing operates. In Spain the awareness among the general public has reached relatively far. There, environmental protection is organised together with local authorities, and local environmental protection groups have been engaged as advisors in order to encourage the staff. This year Sunwing in Spain received a grant from the EU for the education of its employees. The management in Spain has decided that the money will be used for environmental education.

In Greece, on the other hand, the lack of awareness among staff and local authorities is restricting the environmental enforcement. Besides, there is no infrastructure available for the recycling of paper and suppliers offering less harmful products are rarely to be found.

Presently, all hotel managers are instructed to take action based on their individual situation and the conditions prevailing in their respective countries. The written environmental policy of Sunwing, which is dated May 1994, points out choices of materials, economising on resources and energy used and the minimisation of waste and pollution, as guidelines. So far, no steps have been taken to push these ambitions further by a detailed action-plan. In the near future though, a checklist for each hotel (except the hotel in Gambia) will be put together and its usage implemented.

Throughput

Being a service company, the materials usage is relatively low. However, due to the number of guests that Sunwing hosts every year, the flows of matter and energy is significant as well as concentrated at relatively small sites. Regarding the hotels and staff operations, the most significant throughput is the customers. In order to service the customers, the company's operations have the following throughput:

Input:

Water, including water for swimming pools.
Electricity
Foodstuffs for the restaurants and the mini-markets at the hotels (often transported a long way, i.e. from the mainland).
Beverages, wine, beer, mineral-water, soft-drinks and liquor.
Laundry (about half of the hotels have their own laundry facilities).
Paper for computer lists and for communicating with the guests.
Cotton for bedclothes, towels and staff uniforms.
Detergents and chemicals.
Fuel for vehicles and machinery.
Cars and machinery.
Furniture and similar items.

Output:

Waste water (sewage discharges).
Other waste that goes either to landfills or incineration. It is unknown to what extent this part of the waste includes dangerous components such as heavy metals, etc. So far no hotel has started to compost their organic waste.
Paper
Glass
Discharges to the air (for example from vehicles

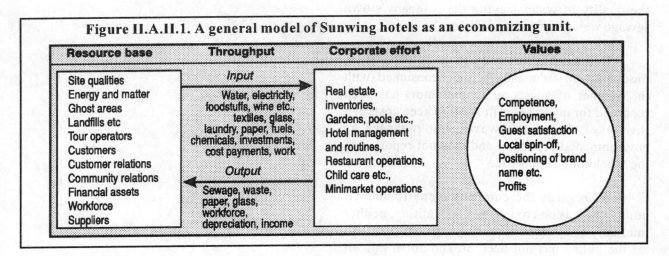

Figure II.A.II.1. A general model of Sunwing hotels as an economizing unit.

Site qualities in the figure refers to land area, land beauty, the beach, the ocean, waste water draining, etc. "Energy & matter" refers to water supply, electricity, foodstuffs, building materials, cotton etc.. In the same way, the rest of the resource base can be specified.

transporting the guests from and to the airport, from rental cars, etc.)

Report routines

Each hotel has a number of standard financial and operational ratios as guidelines. Such ratios are number of occupied beds and rooms, average number of employees, and consumption per guest-night. Every week each hotel reports their liquidity and financial results before taxes to the head-office in Stockholm. On a monthly basis profit and loss account, a balance sheet, a cash flow statement and the above mentioned financial and operational ratios are reported. All reports are sent on forms to Stockholm by mail or by facsimile. All reporting from Sunwing to SLG and from SLG to Airtours takes place with the same frequency, including all the information just mentioned. No information goes the opposite way, i. e. from the head-office and back to the hotels. However, feed-back is given when divergence from the plan or budget is noted.

Environmental reporting

None of the above mentioned reports include any environmental reporting, neither are any reports with regards to the environment delivered to the local authorities. According to Sunwing's controller, Mr. Lars Löfgren, no figures regarding for example waste output are available. Neither are the local authorities requiring Sunwing to report the quality of water leaving the company's own sewage water treatment facilities.

Because it is customary in the hotel business, the quantity of water and electricity consumed (with the number of guests as denominator) has been recorded for many years in order to keep costs at a low level. This is, however, not regarded as environmental reporting and it is not reported to the Stockholm head-office.

As regards the company's environmental policy, there is no environmental auditing, neither internally or by an external auditing consultant. So far the policy has not been broken down into an action plan. A list with all the steps so far taken at the Cala Bona hotel in Majorca (Spain) has recently been put together.

Annex III

SVENSKA McDONALD'S

McDonald's was founded in 1955 in the United States by Mr. Ray Kroc. The idea to start up a chain of hamburger restaurants was derived from two brothers, Dick and Mac McDonald, who successfully sold only one item in their restaurant -- hamburgers. Their restaurant in San Bernadino, California, served as model when Ray Kroc acquired and further developed the brothers' business.

Today, the business-idea is very much the same: a limited and standardized variety of food served in expendable packaging. McDonald's is the world's largest chain of restaurants, with about 14,000 restaurants in 70 countries. World-wide, there is a new McDonald's restaurant opening up every ninth hour.

In Sweden, Svenska McDonald's (by translation: Swedish McDonald's) has had a licence agreement with McDonald's Corporation (USA) since 1973. The company (Svenska McDonald's) has been 100 percent owned by the Lederhausen family until this year when McDonald's Corporation acquired 50% of the shares. The main reason for the family selling a 50 % interest was to strengthen the company for further planned expansion. Today, there are 73 McDonald's restaurants in Sweden: 33 of them are operated by Svenska McDonald's and the remaining 40 are operated by franchise-holders throughout the country.

At the same time as the company pays a percentage of the turnover to the parent-company in the US, the franchise-holders in Sweden pay a percentage of the turnover to Svenska McDonald's. This is the core idea of franchising -- the benefits of well-known brand names and knowledge in return for a percentage of the turnover. In Sweden, all the franchise-holders rent the buildings in which they operate from Svenska McDonald's.

	Svenska McDonald's	Franchise-holders
Ownership (with regards to the buildings/sites)	73	0
Operations	33	40

The company's expansion rate is significant. In 1993, the company opened 9 new restaurants and in 1994 there are plans for even more. The majority of these planned establishments are going to be "drive-thru", or "McDrive" as they are called. (Sweden has the third highest car ownership rate in Europe.)

Because all restaurants in Sweden are owned (with regards to the buildings/sites) by the company coupled with continuous planning, purchasing and construction of more units to come, many of the staff members regard themselves to be working in the real estate business rather than the restaurant business.

To all appearances, the co-operation between the company and the franchise-holders is functioning well. The company's explanation for this is the concept of mutual profitability; both sides are supposed to benefit from the co-operation. From an outsider's point of view the socialization aspect has to be regarded as well, e.g. all managers and owners attend courses at McDonald's University, both in Sweden and in Chicago.

In 1993, McDonald's in Sweden (all 73 restaurants) had more than 60 million guests. The turnover for all these restaurants was SEK 1,345 million. In 1993, the total number of employees (all restaurants) was approximately 7,500 -- the majority part-time workers. For the company's part, the return on equity (SEK 94 million) was 20 percent before taxes that year.

General context

The competition for McDonald's part is principally all kinds of fast-food eating establishments as well as ordinary restaurants serving lunch. Among the chains of hamburger restaurants in Sweden, McDonald's has a market share of 70 percent.

In the last couple of years the disposable incomes among the general public in Sweden has decreased and lunch-eating out at restaurants (which is rather common in Sweden) has become more expensive. This has benefited McDonald's which has responded by not raising their prices.

Business goals and strategy

Svenska McDonald's work is guided by a overriding, verbally-formulated, vision. According to the vision, visiting a McDonald's restaurant should always mean good food, quick service and friendly treatment. Beneath the vision there are three more quantifiable goals:

- 100 percent satisfied customers
- 100 percent satisfied employees
- High profitability

All three of these goals are measured frequently: the two first ones by surveys carried out at the restaurants and the third by traditional means -- return on investment. In order to achieve these goals there are strategies, action plans and sub-goals, both quantitative as well as qualitative. Among the qualitatively-oriented strategies and action-plans, the following can be described:

1. "The four corner-stones"

The original business-idea of Ray Kroc is still valid, internally called "the four corner-stones"; Quality, Service, Cleanliness and Value. The corner-stones speak for themselves, but can be completed by the following: The strategy includes food being delivered only from a few selected Swedish suppliers that have proven able to deliver in accordance with specified standards of quality. McDonald's emphasises profitable and long-term co-operation with all their suppliers. For example,

Svenska McDonald's is still buying all their bread from the same bakery as when the first restaurant opened in 1973. Today, the bakery, which of course has expanded tremendously over the period, operates in the same building as the McDonald's distribution centre in Stockholm.

2. Being a good citizen

One part of the business-strategy is what McDonald's calls "being a good citizen" and "to give something back to society". That means such things as being involved in the local community and employing at least two disabled persons at each restaurant. Another example of this involvement can be mentioned: In 1993, Svenska McDonald's, together with the Children's Cancer foundation, built a house next to a children's hospital in Stockholm. The purpose of the house is to make it possible for the children's families to stay with their children in a home-like atmosphere, yet close to the hospital.

3. Environmental protection

Environmental protection is not one of McDonald's world-wide policies. It can be said to be part of the policy of "being a good citizen", but is not stressed as an imperative policy set by the main-office in Chicago. (It should, however, be noted that there is far-reaching environmental protection work carried out by McDonald's in the US as well as in other parts of the world.)

The environmental protection efforts of Svenska McDonald's were intensified when the former Chief Executive Officer, Paul Lederhausen, met with Karl-Henrik Robèrt in 1991. Mr. Robèrt is the founder of the Natural Step, a Swedish non-government organization stressing the fact that the world's resources are limited and that man-kind has to live in a cyclical society. Co-operation with Natural Step has resulted in the following statement: [Svenska McDonald's long-term goal is] "to be operating in accordance with the natural environment and to be an active part of an ecologically sustainable society".

The first goal set up by McDonald's was a 90 percent reduction of waste within five years --

today they have reached a 65 percent reduction. The present environmental policy goes further than just waste reduction as it states that the use of energy and natural resources in the future shall be minimised and that all parts of the operations shall be in accordance with "the principle of cyclical society", i.e. a sustainable society. The first goal in order to achieve this has been to educate all managers. The next step will be to educate all 7,500 employees with the help of a multimedia-device that is now under development. Further, all suppliers, as well as consultants and constructing contractors, are required to participate in courses in environmentally-sound thinking. The plans for the future have been divided into the following 6 groups:

- Food. Within the next months, three of the restaurants will try to make compost from all of their organic waste. The cups containing ice-cream (that used to be made of plastic) are now made of eatable biscuit and the spoons are made of wood -- thereby being compostable.

- Distribution. The distribution of all food is very centralised with only one supplier of each foodstuff (bread, meat, french fries, lettuce, etc.). Most of the food is transported to Stockholm and redistributed daily by McInco, Svenska McDonald's own transport company. Svenska McDonald's states in their action plan that the amount of railroad transportation is going to be increased and that they will try out the use of bio-fuels instead of fossil fuels.

- Waste. Here the efforts include sorting out items at the source, recycling and substitution of plastic materials for renewable materials, where possible.

- Cleaning and chemicals. Purchasing biologically degradable detergents, etc. from suppliers.

- Office equipment. The best available materials must be used and old computers are to be recycled.

- Construction. Phasing out the most harmful materials. Ecologically-sound buildings and the use of natural materials are strived for.

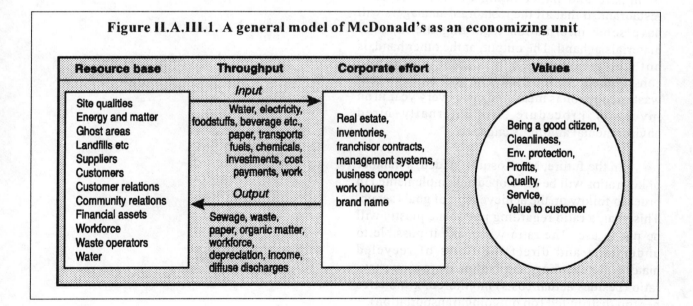

Figure II.A.III.1. A general model of McDonald's as an economizing unit

Resource base	Throughput	Corporate effort	Values
Site qualities Energy and matter Ghost areas Landfills etc Suppliers Customers Customer relations Community relations Financial assets Workforce Waste operators Water	*Input* Water, electricity, foodstuffs, beverage etc., paper, transports fuels, chemicals, investments, cost payments, work *Output* Sewage, waste, paper, organic matter, workforce, depreciation, income, diffuse discharges	Real estate, inventories, franchisor contracts, management systems, business concept work hours brand name	Being a good citizen, Cleanliness, Env. protection, Profits, Quality, Service, Value to customer

Report routines etc.

All the restaurants, no matter if they are owned or operated by franchise-holders, are reporting their daily sales figures to Svenska McDonald's head-office in Stockholm. Every week a more detailed report is prepared: the report includes figures representing turn-over, etc., as well as performance indicators such as number of guests, sales per guest, and cost of food in proportion to total cost. (Actually, all financial reporting follows the same standards all over the world.) The personnel at the head-office in Stockholm put together and return comparable sales-reports to all the restaurants. The information is supplemented with figures regarding deliveries, for example number of bread-buns, kilograms of french fries, etc.

The environmental manager, Bertil Rosquist, says that they are presently having the reporting system improved by the soft-ware supplier (at a significant cost) in order to add two more columns to the reports. The columns will include kilograms of plastic and paper being delivered to each restaurant so that all the restaurant-managers will have some information regarding the input of materials at hand. The output, at the other hand, is not yet measured on a regular basis, and consequently not reported. The goal of 90 percent waste-reduction is measured once every year in an inventory procedure. No externally hired environmental auditor is engaged.

In the future, Mr. Rosquist added, a number of key ratios will be developed and implemented in order to follow up the achievement of goals set up. This year, a ratio regarding polythene plastic will be put in use. The ratio will make it possible to understand and direct the flows of recycled materials. Regarding distribution, there is one key-ratio (besides actual costs) monitored: kilometres per ton distributed (on outgoing transportation).

It has been made obvious to the management that significant changes in output of waste occur when Svenska McDonald's changes something in the design of the packaging. For example, the cardboard box for french fries has been substituted with ordinary paper. That change immediately resulted in a noticeable difference compared to the differences that changes in processes and habits at the restaurants usually contribute to.

Regarding the qualitative goals that account for all restaurants in the whole world, there is no reporting from Sweden to McDonald's Corporation by traditional means. However, the field-service department, consisting of 6 persons who previously were employed by McDonald's Corporation (USA), are monitoring all aspects of the fundamentals in the McDonald's concepts that are followed. Since the goals in question are not easily measurable, these persons are highly dependent of their intuitive feelings for what is appropriate or not.

Because of the company's goal to meet the prerequisites of a cyclical society, there is currently a researcher at the University of Lund engaged in a project which aims to try out an expanded accounting model. In the future, three kinds of capital might be reported; financial, natural and human.

Annex IV

VOLVO AB

History and market situation

AB Volvo, one of Sweden's largest companies, produces and markets engines and transportation vehicles. In 1915, the company was formed as a subsidiary of the Swedish ball-bearing manufacturer, SKF, and by 1932 was assembling automobiles, trucks, and bus chassis in Göteborg. Today, the company's main emphasis is still autos and trucks, even after diversifying in the 1980s into the drug, food, and energy sectors. Businesses outside of the core transportation industry have been either sold or become less important. Also the proposed merger of Volvo's automotive operations with the French manufacturer, Renault, was withdrawn by Volvo's Board of Directors. Volvo is now, more than before, a autonomous car company where the sales from the car group and the trucks group account for 85% of Volvo's total sales.

In 1993, there were 147,300 shareholders. However, the 250 largest shareholders owned over 75% of the company. These owners are mainly funds and insurance groups.

By the end of 1993, Volvo was active in most of the world's markets. Volvo's largest market is the USA. Great Britain, Sweden and Germany are also very important markets for the company and these four markets together represent more than 50% of total sales. Currently, the fastest growing markets for Volvo are in the Far East and eastern Europe, but they are still small in comparison to the traditional markets.

Sweden is unique in that this small and sparsely populated country has two automobile manufacturers--Volvo and SAAB. Volvo is the larger of the two; however, the two are active not only in the same markets, but also in the same segments of those markets.

In 1993, the trend was that the demand for new cars in western Europe decreased and this affected Volvo. The car industry is also known to have had problems with over- capacity. Decreasing demand and over-capacity give a picture of a competitive market situation in which the car companies more than ever need to have a business concept that gives give their products a clear image.

The Volvo core values and guiding stars for long term growth and profitability, as stated in the annual report, are:

- Quality
- Safety
- Concern for customers and the environment

Volvo and the environment

Volvo's first environmental policy originated in 1971. It was produced in connection with the first UN environmental conference and stated that cars should not be built at any environmental cost and that the environmental issues were of importance to Volvo. In 1988, a working group of senior managers from all areas of the company produced a new environmental policy. The idea was that the policy should be based on a holistic approach to the environmental impact of Volvo's products. In 1989, it was introduced throughout the company. The policy pledges:

1) to develop and market products with superior environmental properties and which meet the highest efficiency requirements;
2) to use manufacturing processes that have the least possible impact on the environment;
3) to actively participate in and conduct their own research and development in the environmental field;
4) to select environmentally compatible and recyclable material in connection with the development and manufacture of products, and when purchasing components from suppliers;
5) to apply a total view regarding the adverse impact of products on the environment;
6) to strive to attain a uniform, worldwide environmental standard for processes and products.

In Volvo, every company manager is responsible for implementing action consistent with the policy.

Volvo's environmental council has the final responsibility for the policy. The Council is comprised of specialists from the Group's head offices and from the various member companies, and the organization is responsible for the ongoing co-ordination of activities. The Council also oversees an environmental audit. The policy, although fixed in its theoretical approach, allows modifications when necessary to adjust to new environmental findings and problems. In a new version of the policy, suppliers doing business with Volvo must also comply with environmental requirements.

Volvo argues in the written information available that an effective environmental effort is the basis for long-range profitability and favourable economic growth. This, in turn, is said to promote rational, ecologically-oriented investments in production and products. Production and products go hand in hand and Mr. Inge Horkeby, a member of Volvo's environmental council, says that "the cleanest car cannot be produced in the dirtiest plant".

Resource-base

There has not been any war on Swedish soil in over 300 years, and historically, Sweden has had political stability. Sweden has also had, for the last forty years, a rather stable economy, and the Swedish tax regulations have, to some extent, levelled out differences in income. The Swedish family has had the financial possibilities to own a car and the Swedish car ownership per capita ratio is among the highest in the world.

Sweden is a democracy with a market economy. and this has made it possible for other producers to set up business and compete with Volvo. There has been both a high degree of demand, as of supply, concerning the products that Volvo produces.

The cost of industrial labour in Sweden has been high compared to other European countries, and this has contributed to a high degree of automation. Over 90% of the nation's youth go to high school, resulting in a highly-educated populace.

Sweden has 8.5 million inhabitants, averaging 21 persons per square kilometer. Sparsely populated, most inhabitants live in the southern part of the country where most industries are located. The northern part of Sweden has a number of natural resources that are important for industry - - iron ore, water power, and forestry. Not only are these raw-materials important for the production of vehicles but the exploitation of natural resources also requires transportation.

The Social Democrats, known to be supportive of large corporations, have been the most influential political party in the Swedish Parliament during this century. The car industry needs roads and highways and the needed infra-structure has been financed by the tax-payers.

Volvo's importance to the Swedish economy cannot be minimized. Volvo's production depends on supplies from smaller firms throughout Sweden, and because of this, Volvo plays an important political part in the Swedish transportation industry.

All these things put together have helped Volvo to achieve a brand-name confidence, and, thus, a 25% market share in Sweden.

Operations

Volvo's operations are basically divided into five areas. known as different Groups. The Groups consist of:

- Volvo Car Group
- Volvo Truck Group
- Volvo Penta Group (marine engines)
- Volvo Flygmotor Group (aircraft engines)
- BCP Group (consumer goods)

The organization is decentralized and there are many companies and profit centers in every Group.

Volvo is one of the three largest truck manufacturers in the world, producing 50,900 vehicles in 1993. The market share remained unchanged at 10% in 1993. The Volvo Car Group is not as large in comparison with its competitors. Automobile production in 1993 was 290,700 units, with 75% of the autos being produced outside of Sweden. In 1993, the number of cars sold in the domestic market was 12.5% of Volvo's total sales.

In 1993, Volvo employed 73,641 people world-wide. Since 1991, the number of domestic employees has decreased by 6,000 (14%), and in 1993 there was almost an equal number of domestic and international employees. The plants are situated all over the world and Volvos are being in made in e.g. Sweden, Belgium, Canada, Thailand, Malaysia, and Scotland. Because of this, Volvo is dependent not only upon trans-national relations, but also on trans-national environmental issues.

Throughput

Motorism in Sweden as well as in many other countries has been pointed out as one of the main environmental problems. Volvo's concern, from this point of view, is not only the manufacturing of the products but also the products themselves. The environmental impact of the operations are substantial and Volvo needs for legal, market and value-reasons to deal with the problem. The company describes its environmental view as holistic. The environmental policy is to look at the impact when natural resources and raw materials are used, when producing the products, when the product is in use and when the car or truck is disposed of. Volvo uses many different kinds of raw materials and chemicals in its production and Mr. Horkeby says they want to promote "good examples". The paint shop in Umeå is a typically good example, says Mr. Horkeby, since it can boast as being in the lowest emissions level in its sector.

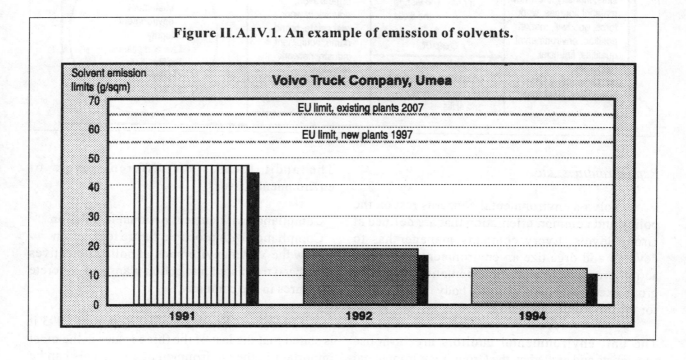

Figure II.A.IV.1. An example of emission of solvents.

In a company the size of Volvo, the use of human resources is substantial. Volvo needs many different kinds of qualified persons to develop, produce and market its products. Volvo needs to hire competent people and develop them throughout the time they are active in the company. They also need to retain their personnel so that the labor turnover doesn't deplete the company of the human resources it requires to meet and maintain its explicit (and implicit) values.

Values

Volvo is listed on several stock exchanges and profitability is thus a main goal.

In 1993, operating profited (1,548 MSEK) after two years of losses. However, the return on shareholder's equity in 1993 still showed a loss. Volvo has a profit-related bonus system that during the 1988-1993 period distributed a total of 1,403 MSEK to Volvo employees.

Volvo's management argues that profitability, as was noted above, can be reached by producing a car whose characteristics are, quality, safety and environmental concern. These attributes can be seen as values of their own.

In 1992, Volvo produced its Environmental Concept Car (ECC). It is powered by an electric/ gas turbine and Volvo calls it "a realistic option for the family car of the future".

A general economic model of the company is showed below in figure II.A.IV.2.

Figure II.A.IV.2. The general structure of AB Volvo shown in an economic model concerning the circulation of resources.

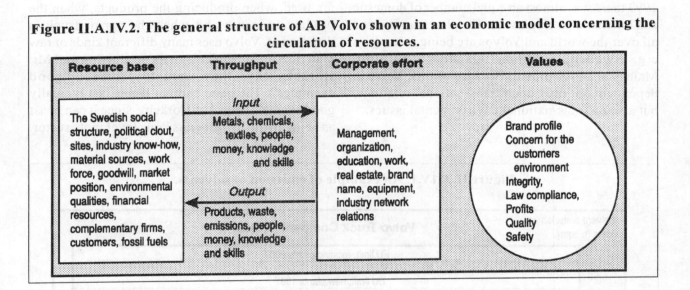

Report routines, etc.

Volvo's environmental concerns rest on the policy and common orientation that are defined at Group levels. Every company manager has to develop and organize an environmental strategy, action programme and implementation plan. The group executive board's control body is responsible for the environmental audit.

The unit environmental auditors are in-house specialists and are led by the Group's environmental auditor. The reason is that auditing needs open records with no hidden items. The auditors are always from outside of the unit being reviewed. Only on rare occasions have the auditors been from outside Volvo. This happens when special legal knowledge is needed.

The audit focuses on different levels of environmental issues:

_ Compliance with current national legislation
_ Compliance with probable future legislation
_ How the Volvo environmental policy objectives have been translated into action plans and concrete measures in the company.

In Volvo's Environmental Report of 1993 it is stated that the last of the three issues is the most important in the environmental audit. This can be a problem when dealing with different cultures, different views of the environmental problems and different legislation, says Mr. Horkeby. The limiting factor is legislation and at times this is where the environmental work ends.

The integration of environmental performance indicators
with financial information by transnational corporations

75

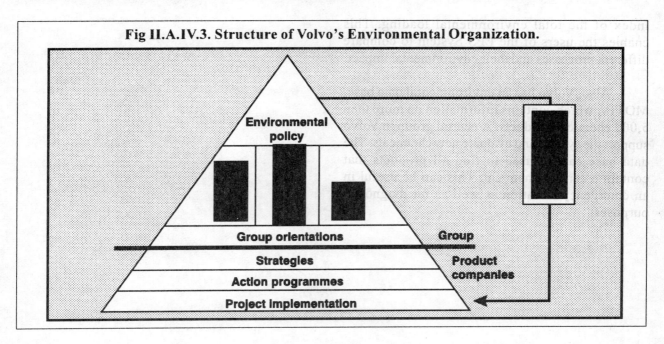

Fig II.A.IV.3. Structure of Volvo's Environmental Organization.

Critics of the company argue that Volvo only acts under pressure and that the company consequently refrains or postpones investments that could improve its environmental performance. The reason is, says the critics, that the investments necessary for environmental improvements are seen as too expensive. There seems to be a conflict between profitability and environmental performance.

Volvo's management is very proud of the decentralized structure of its organization. The policy, on the other hand, is a centralized product. This could be another reason why it can be hard to get the individual operating units to observe the policies in daily operations.

The company does not perform a yearly environmental audit. It depends on surprise audits to get an accurate reading on policy compliance. A letter is sent by the auditors to the Group's environmental manger and managing director requesting the necessary information. Usually this information is found in applications for concessions requested of outside regulatory agencies such as the National Franchise Board of Environmental Protection. Other sources of information could be blueprints concerning sewers, environmental action plans, education plans, environmental reports, and purchase lists for raw materials and chemicals. The written information is combined with on-site

interviews with personnel. An exit report is produced on the basis of these data, and the report is sent to all appropriate managers, as well as to the Group Executive Board. The audit is more than financial auditing: it is more of an evaluation. Mr. Horkeby emphasizes that the audit gives no guidance as to how the company should improve its environmental actions. The auditors only point out where poor environmental performance can be found and it is the responsibility of the company managers to take appropriate measures.

Tools for environmental action

Volvo produces a separate Environmental Report and also has an environmental news letter within the company. More systematized information is available for the product developers in the so-called EPS System (Environmental Priority Strategies in product design). It was developed by the car industry together with the Swedish Environmental Institute (IVL) and the Federation of Swedish Industries. EPS is a life-cycle analysis system that helps the Volvo companies to examine the environmental impact of materials and products. The key-element of the EPS-system is an environmental load index that is a weight of the environmental impact of items such as each chemical. Different weights are produced by IVL. Another feature of the system is that different factors are weighted together to produce a single index of the total environmental loading. This enables the users of the EPS System to compare different processes and their environmental impact.

Also, Volvo has an environmental data base, MOTIV, which contains information on more than 5,000 chemical products. A central group in Volvo support the system and evaluate new chemicals. The data base can be used to find all products that contain a certain chemical. This can be useful in an emergency situation as well as for diagnostic purposes.

Annex V

STORA SKOG

The STORA group is a Swedish multinational company based on raw materials, including: forests, energy, pulp and paper, construction material and packaging. It is also one of the oldest companies in the world since, in 1992, it celebrated it's 700 anniversary. The STORA group is today one of the major forestry enterprises in Europe with annual sales in 1993 exceeding 50 billion SEK and with more than 33,000 employees, mainly in Sweden and Germany.

STORA Skog is the forestry division within the STORA group. In 1993 the company's annual sales were 5.7 billion SEK, and it had 2,400 employees. STORA Skog administrates 1.6 million hectares of productive forests, mainly in the central part of Sweden, and 0.8 million hectares outside of Sweden. Annually STORA Skog supply 13 million cubic meters of timber to its own sawmills and pulpmills, one third cut in its own forests. From its four sawmills in Sweden with an annual capacity of almost 700,000 cubic meters, nearly 80% of the production is exported, mainly to The Netherlands, Great Britain, France, Denmark and Norway. (source: annual report of The STORA group, 1993)

STORA Skog recognized at an early date the importance of environmental consciousness. This started primarily as a way of getting more freedom of action in relation to the increasing political and legal concerns on environmental questions: instead of being in the position of only responding to new legal regulations, STORA Skog tried to set standards higher than the actual minimum requirements. Although initiatives directed towards forestry companies have also come from environmentalist non-governmental organizations such as Greenpeace and the WWF, concerning the use of tropical timber, sub-alpine forests, primeval forests and clear-cuttings, this in turn raised a market demand to monitor the origin of the used raw material, e.g., chain of custody. Today STORA Skog and the other major forest companies in Sweden are also having serious discussions with Greenpeace and other non-governmental organizations on a more or less regular basis.

In an interview with the Chief Executive Officer of STORA Skog, Professor Björn Hägglund, he pointed out some initiatives that have already been taken by STORA Skog. In Sweden STORA Skog was the first company to stop harvesting in sub-alpine coniferous forests in northern Sweden. Today there is a current moratorium, but if this should be lifted, harvesting in sub-alpine forests will be preceded by a special inventory and consultations with representatives of nature conservation organizations. Also, conflicts between the Samic population and forest companies in northern Sweden were first settled by STORA Skog together with other companies such as MoDo and Korsnäs.

A more symbolic issue is the preservation of the white-backed woodpecker, a bird with almost the same symbolic value as the northern-spotted owl in the United States. In the areas where STORA Skog has its forests there are less than 50 pairs of white-backed woodpeckers alive, and still this is almost the entire Swedish population of the bird. Instead of being forced to preserve the bird, STORA Skog voluntary decided to set aside hundreds of small sanctuaries in order to save the remaining population.

In 1993 STORA Skog adopted a forest management strategy which aims to sustain the forests both as a source of timber and as a habitat for flora and fauna. The policy is to maintain the biodiversity of forest land and at the same time increase both the production of timber and the economic value of the forests. The plan is based on ongoing research both in silviculture and conservation.

The aim for STORA Skog's forestry operations from 1993 onwards is to contribute towards the achievement of the STORA groups financial objectives -- and at the same time secure a high, valuable and sustainable yield of forest production whilst maintaining the biodiversity of the forest land. (source: Diversity for the Future, brochure prepared by STORA Skog, 1993)

In order to achieve this, four principal objectives have been defined:

* *To increase the annual cut in the long term.*

This objective is based on the fact that today the annual cut is only 80% of the annual increment in the forests. This means that the total volume of standing timber is increasing each year in forests owned or controlled by STORA Skog, therefore sound practices will enable STORA Skog to obtain a higher yield.

* *To produce high-grade sawlogs from the pine stands and to produce high-grade raw materials for fiber products from the spruce stands.*

The conditions in Scandinavian coniferous forests are especially conductive to long fibre species of trees. Hardwood will not be planted as a commercial crop, but only to maintain biodiversity.

* *To preserve the biodiversity of the forest landscape.*

STORA Skog have devised a strategy for conservation at the regional level. It is based on ecological landscape planning: this means that, in a given region, a policy of planned conservation is undertaken. Mixed forest-management methods are to be used: for instance by setting aside certain trees, or by the controlled burning of certain forest areas. Conservation in daily forestry operations: for instance by setting aside individual trees from harvesting, or leaving undisturbed habitats alongside lakes, watercourses, and marshy land.

The STORA Skog forest-management strategy was approved at board level meetings, both at the STORA Skog group level and at the parent company STORA group level. The decision includes a time-limit whereby the landscape planning programme should be accomplished within 10 years, e.g., by the year 2003.

From a resource circulation point of view, STORA Skog has a long tradition of studying differences in growth, quality, and possible output under various inputs. The aim is to find a sustainable balance between cutting, planting, fertilizing and environmental concerns. Similar programmes can be found among most of the other major forest industries. In general, forest companies consider themselves to be one of the most sustainable businesses since the operations are based on closed circulations.

On an operating unit level, STORA Skog has not adopted any life-cycle analyses concerning matters such as: energy consumption or amount of fertilizers used per acre or per cubic meters. The transportation with trucks and the heavy mechanization of the cutting in the forests is a crucial problem from an environmental point of view. The heavy concentration on mechanization is an obvious consequence of the reduction of employees from some 15,000 in 1980 to 1,500 working today in the forests . However, the machinery used now operates using more environmentally-friendly diesel and vegetable oils, and the fertilizers used are less hazardous.

A discussion of external certification and environmental auditing is on the agenda at STORA Skog. Today STORA Skog has no external environmental auditing; instead this is accomplished by its own personnel. The question of third-party certification is also a key problem at the moment. The idea is not so much about just certifying the final product; instead, it is to get a neutral third-party certification of the whole process, from the tree in the forest to the final product. This is an obvious problem for companies such as STORA Skog which might have thousands of suppliers to one single pulpmill. Some of the raw materials are also traded and exchanged prior to the arrival at the pulpmill. The Chief Executive Officer stresses the importance of enjoying public confidence in how the forestry industry is managed in general, and especially from an environmental point of view.

A public environmental audit report for 1994 will be prepared in 1995. This report will include figures about the general standard of the forests, as well as ratios of annual cutting volumes compared to the annual growth of the remaining tree population.

The question of how to integrate financial and environmental reporting is an extremely difficult one for a forestry company such as

STORA Skog. In a general financial report one important question is the value of the stock. The stock for a forest company is the forests. To put a monetary stock-value on a forest includes a difficult discounting problem when the net present value of a forest, which might have an average length of life of 80 years from now, is to be developed. This is practically impossible due to all the uncertainties involved in such a calculation. From all appearances, no Life Cycle Analyses or regular environmental on-site indexes are calculated, and this implies that environmental and financial reporting are not being presented together on a regular basis.

STORA Skog's Chief Executive Officer especially emphasized the need to develop and use information technology in the forestry industry. This will improve the environmental monitoring possibilities. Landscape-planning is also very much a question of communication and information as to how the forest on a local level will be developed under different circumstances.

Also the increasing demand to monitor the whole process from tree to final product is very dependent on reliable information. And finally, the whole question of integrating financial and environmental information is yet to be solved. In the near future there will be an increasing need for standardized, reliable and confident methods as to how to integrate environmental information with existing financial information.

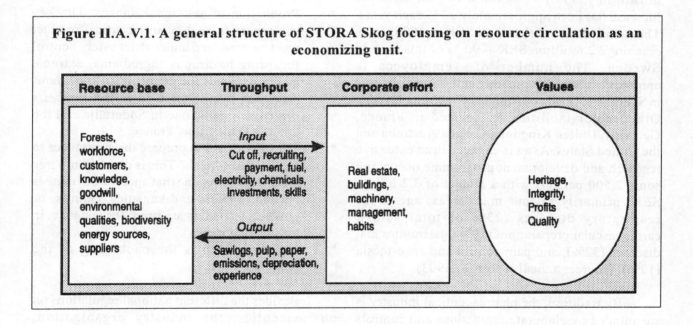

Figure II.A.V.1. A general structure of STORA Skog focusing on resource circulation as an economizing unit.

Annex VI

ASTRA

Astra is a Swedish multinational company in the pharmaceutical industry that was founded in 1913. The head office is located in Södertälje, Sweden, which is 25 kilometers south of central Stockholm. In the early 1930s Astra began a research and development programme that has introduced pharmaceuticals such as: *Xylocain* (anaesthesia), *Seloken* and *Plendil* (hypertension), *Bricanyl* and *Pulmicort* (asthma), *Penglobe* (penicillin), and *Losec* (gastric ulcer). Today Astra is one of the fastest growing pharmaceutical industries in the world with an annual volume growth in 1993 of 45 % (on a 10-year basis the increase has been approximately 20 % each year). The net profit was 6.1 billion SEK in 1993; sales reaching 22.6 billion SEK -- 90 % of this outside Sweden. The number of employees is approximately 12,800, mainly at the Swedish units in Södertälje, Lund and Mölndal (7 800 employees). Other major facilities are located in France, Germany, United Kingdom, Canada Australia and the United States. Astra is engaged in an extensive research and development programme occupying some 2,500 persons with a budget of 3.2 billion SEK, primarily in four main areas: agents for respiratory diseases (23% of total sales), cardiovascular preparations (19%), gastrointestinal diseases (32%), and pain control and anaesthesia (12%). (source: annual report for 1993)

By tradition, the pharmaceutical industry is surrounded by elaborate regulations and controls prior to the release of a product to the market. Primarily this concerns testing, documentation and clinical examinations. A patent for a product is usually valid for twenty years, and due to significant competition from other pharmaceutical companies and heavy investment demands, a new product or process is patented as early as possible. This means that as much as ten to fifteen of a patent's twenty year duration is used up prior to market release. Therefore the remaining five to ten years has to generate a positive cash-flow that is sufficient to cover the initial investments and generate substantial profits. Astra has also followed the rule

that all research should be financed from internal sources without loans, due to the financial risks involved in long-term research operations.

Astra is at the moment the world's second largest manufacturer (in sales figures) of gastric ulcer pharmaceuticals, the third largest for asthma treatment, and is leading the industry in local anaesthetics with almost 70% of the world market.

The pharmaceutical production is basically divided into three phases:

*   Production of active substances. The key substances in each pharmaceutical product must be produced under strict safety control regarding hazardous ingredients, extreme quality demands and protection of know-how. Astra has two of these so-called *synthesis production* units, one in Södertälje and the other in Dunkerque, France.
*   The next step is to prepare the substances to be pharmaceuticals. This is divided into tree areas depending on what specific attribute is needed. Tablets and capsules, liquids, or powders (powders are used, for instance, in asthma treatments).
*   The final step is the packaging of the pharmaceutical

Besides the different national regulations on pharmaceutics, the industry organization, Federation of Pharmaceutical Manufactures Associations (IFPMA), has a code of conduct and sets standards, for instance with regards to marketing.

The external environmental questions were first operationalized by the company in 1965 when an environmental department was formed. This work became even more intensified in 1968 when a so-called "emission group" was established. The purpose of this working group is to study the effects and consequences on water and air pollution caused by the manufacturing of pharmaceuticals. The ongoing work has since intensified, and from 1983 the environmental department -- *Environmental Affairs* -- has been a separate unit in a group called Manufacturing and Logistics Support. The Manufacturing and Logistics Support group in Sweden operates directly under the board of directors and is lead by a Vice President, Mr. Jan Larsson, although the President and Chief Executive Officer, Dr. Håkan Mogren, has the final responsibility. Dr. Mogren also initiated and signed the present environmental policy (1991) and receives all major reports on environmental concerns.

The Environmental Affairs department today has seventeen persons working with internal and external environmental issues and chemical and occupational matters co-ordination both within manufacturing as well as in research and development. This also includes an environmental engineering group monitoring, for example, the sewage water from Astra's plants in Södertälje while a new sewage water plant is under construction. Within the department groups are also monitoring hazardous raw-materials and wastes, co-ordinating environmental training, managing external relations and concessions etc.

To reach the high standards set by Astra, all production is strictly supervised, which provides an opportunity to monitor environmental impacts. Within Astra the quality policy is formulated in a programme called: *Quality Assurance Guidelines*, which is a part of the external so-called "GMP" regulations (*Good Manufacturing Practice*). Due to the risks involved in manufacturing with hazardous components, risk assessments are conducted on a regular basis. The frequency of these assessments depends on the classification of the various substances. In general, a precautionary principle prevails: to use as much closed production systems as possible, substitute less hazardous components whenever possible, and clean, reuse and recycle.

Two examples of these substitutions are:

* To discontinue the use of CFCs and HCFCs. Astra is switching over from CFCs/HCFCs in cooling systems to ammonia, liquid nitrogen and hydrocarbons. The present emission of CFCs/HCFCs from Astra's units in Södertälje is less than 2 metric tons each year.
* To significantly reduce the emissions of chlorinated solvents. At Astra's Södertälje units the amount of chlorinated solvents discharged has been reduced from 250 metric tons in 1984, to 33 metric tons in 1993 (87% reduction). At the same time the level of other solvents discharged has been reduced from 400 metric tons in 1984 to 300 metric tons in 1993 (25% reduction).

However both of these reductions were primarily initiated by outside pressures and forthcoming moratoriums on production and purchases.

The introduction of the *Turbuhaler* inhalator used for asthma treatment has eliminated the use of aerosols as propellants. Instead, the inhaler is driven solely by the patient's deep breath. The use of *Turbuhalers* instead of aerosols has increased from one million units in 1989 to fifteen million units in 1993. The total reduction of world aerosol consumption from this product was approximately 500 metric tons between 1989 and 1993. The *Turbuhaler* is also made of chlorine-free plastic that can be burnt after use. A study is also being conducted to determine if the *Turbuhaler* can be manufactured so as to be recyclable.

To promote the reduction, reuse and recycling of the packaging for pharmaceuticals, an overall environmental policy is in process of preparation.

The environmental work carried out by Astra has started in countries where the legal framework concerning environmental considerations has been the most developed. The units in Sweden, western Europe and North America have been the first to accomplish new standards. However the environmental policy which was adopted in 1991 and which is valid in all countries where Astra is present has not been implemented as yet at all company locations. Every unit manager is responsible for integrating the environmental policies in daily operations and to establish ongoing three-year plans that include local concessions. Also more long-ranged plans are being developed. For example, at Astra's Södertälje units an action plan reaching to the turn of the century will be finished in October 1994. Astra in Sweden is also carrying out optional programmes such as mass balances which monitors the use of raw materials, emissions to bodies of water and to the air, and solid wastes from specific projects or production lines.

In Sweden the legal system also requires an Environmental Impact Assessment (EIA, "MKB", in Swedish) for all changes or extensions of present business operations. The Swedish EIA is analogous to the Seveso directives for the pharmaceutical industry within the EC countries in terms of risk analysis. The expansion of the Astra unit in Dunkerque, France is in compliance with these Seveso directives.

Environmental auditing at Astra is primarily conducted by internal personnel who visit the different units. However at Astra's units in Södertälje, auditing is led by an external auditor. An environmental auditing report was summarised in the 1993 annual report, but from 1994 onwards

it will be presented as a separate report. An internal environmental report is presented to the board of directors whenever any important change in production is taken -- otherwise the board of directors receives no formal environmental reports on a regular basis besides the annual reports.

Key figures concerning data on significant environmental impacts are developed on a regular basis at each unit. These figures monitor the impact levels stated in concessions and regulations as well as internal goals set up at each specific unit. The frequency of monitoring depends on the level of importance from environmental impact point of view, as well as on its ability to be measured. In Södertälje five or six different figures are monitored on a daily basis, for instance: heavy metals, chlorinated and non chlorinated solvents, different parameters on waste water (COD, BOD, solvents, heavy metals, toxicity, etc.,) and the use of raw materials.

However these key-figures are not related to financial factors -- i.e., there is no integrated environmental and financial reporting.

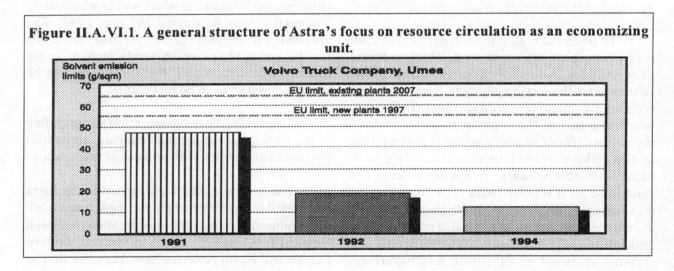

Figure II.A.VI.1. A general structure of Astra's focus on resource circulation as an economizing unit.

In an interview with Mr. Lars-Göran Bergquist, Director of Environmental Affairs at the Manufacturing and Logistics Support unit, it was emphasised that the pharmaceutical industry is operating under extremely regulated circumstances regarding testing, documentation, safety and quality. The pharmaceutical industry cannot afford,

or be allowed, to be innovative to the same extent as manufacturers of consumer goods, such as the producers of household washing detergents. Therefore the whole pharmaceutical industry can be characterised as law-driven and conservative. Even if there is a lot of innovative persons involved in research and development, it still takes some ten years before the innovations will reach the market. Any changes in the content of a drug requires new testing and license application. The characterisation of the industry as law-driven and conservative is to some extent also valid for the measures taken from an environmental protection point of view -- these measures are often initiated by governmental authorities. The potential for increasing environmental awareness for pharmaceuticals is more focused on improved production processes and new packaging. The US regulations have especially affected the industry, particularly regarding a patient's safety. Mr. Bergquist is also expecting, for example, that the US environmental assessment programmes soon will be implemented in the western Europe including Sweden. However Mr. Bergquist did not foresee any immediate integration of environmental and financial reporting, even though, key-figures are currently available.

References

Andrews, P.W.S., 1949, <u>Manufacturing Business</u>. London: MacMillan.

Bateson, G., 1972, <u>Steps to an Ecology of Mind</u>. New York: Ballantine Books.

Bergström, S., 1992, "Ecology, Economy and Value Theory." In <u>Human Responsibility and Global Change</u>, ed. by L.O. Hansson and B. Jungen. Heidelberg: Springer.

Bergström, S., 1993, "Value Standards in Sub-Sustainable Development. On Limits to Ecological Economics." <u>Ecological Economics</u> 7:1 pp. 1-18.

Bergström, S., 1995, <u>Sustainable Management</u>. On Discretion between Endowment and Quality. New York: Island Press (forthcoming).

Björsell, M., 1993, <u>Environmental Expenditures in the Swedish Manufacturing Industries</u>. Stockholm: Statistics Sweden/Env. Statistics Programme. August 1993.

Buzzel, R.D., and B.T. Gale, 1987, <u>The PIMS Principles - Linking Strategy to Performance</u>. New York: The Free Press.

Chandler Jr., A.D., 1977, <u>The Visible Hand</u>, Cambridge, Mass.: Harvard University Press.

Common, M., 1988, "Poverty and Progress Revisited". In <u>Economics, Growth and Sustainable Environments</u>. David Collard et al (eds.), London: Macmillan, pp. 15-39.

Cook, E., 1976, <u>Man, Energy, Society</u>. San Francisco: Freeman.

Cyert, R., and J. March, 1963, <u>A Behavioral Theory of the Firm</u>. Englewood Cliffs, N.J.: McGraw-Hill.

Daly, H., 1977, <u>Steady State Economics</u>. San Francisco: Freeman.

Douglas, M., 1986, <u>How Institutions Think</u>. Syracuse, N.Y.: Syracuse University Press.

Drucker, P.F., 1993, <u>Post-Capitalist Society</u>. New York: Harper Business.

Earl, P., 1983, <u>The Economic Imagination</u>, New York: Sharpe.

Eisner, R., 1988, "Extended Accounts for National Income and Product", <u>Journal of Economic Literature</u>, Vol. XXVI (Dec. 1988) pp. 1611-1684.

Harris, M., 1977, <u>Cannibals and Kings: The Origins of Culture</u>. New York: Random House.

Johnson, H.T., and R.S. Kaplan, 1987, <u>Relevance Lost: The Rise and Fall of Managerial Accounting</u>. Cambridge, Mass.: Harvard University Press.

Lovelock, J.E., 1982, <u>Gaia, a New Look at Life on Earth</u>. New York and Toronto: Oxford University Press.

The integration of environmental performance indicators
with financial information by transnational corporations

85

Miller, J.G., 1978, Living Systems. New York: McGraw-Hill.

Myrdal, G., 1975, Against the Stream, New York: Vintage Books.

Nordhaus, W., and Tobin, 1973, "Is Growth Obsolete?", in The Measurement of Economic and Social Performance, ed. by M. Moss. New York: National Bureau of Economic Research.

Prigogine, I., and I. Stengers, 1984, Order Out of Chaos: Man's New Dialogue with Nature. New York: Bantam Books.

Ryding, S-O., and B. Steen, 1991, EPS-systemet, IVL, Göteborg, B 1022.

Standard & Poor´s, 1992, International Criteria. New York (June).

United Nations, 1990, Information Disclosure Relating to Environmental Measures, New York: United Nations.

United Nations, 1991, Accounting for Environmental Protection Measures, New York: United Nations.

United Nations, 1992, Environmental Disclosures: International Survey of Corporate Reporting Practices, New York: United Nations.

United Nations, 1993, Environmental Management in Transnational Corporations: Report on the Benchmark Corporate Environmental Survey. New York: United Nations, 1993, Case Study in Accounting for Sustainable Forestry Management, New York: United Nations.

Vester, F., 1988, Leitmotiv vernetztes Denken. Für einen besseren Umgang mit der Welt: Munich: Wilhelm Heyne Verlag.

Welford, R., and A. Gouldson, 1993, Environmental Management and Business Strategy: London: Pitman Publishing.

Young, M.D., 1992, Sustainable Investment and Resource Use: Man and the Biosphere series, vol.9. The Parthenon Publishing Group, London: UNESCO.

Zolotas, X., 1981, Economic Growth and Declining Social Welfare. Athens: Bank of Greece.

Notes

[1]   Technically, the format of margin measures varies, depending on different structures in resource bases.

[2]   E/C.10/AC.3/1992/3, p.2.

[3]   Sales No. E.94.II.a.2, pp. 75 and 171.

[4]   Ibid., p. 1.

[5]   Ibid., p. 168.

Chapter III

REVIEW OF NATIONAL ENVIRONMENTAL ACCOUNTING LAWS AND REGULATIONS

Report by the UNCTAD secretariat

Summary

This report reviews the legal requirements that exist in United Nations member States on environmental accounting matters. It is based upon replies to a questionnaire sent to all member States.

INTRODUCTION

This report is based upon the replies to a note verbale and questionnaire circulated to all member States. Additionally, the questionnaire was sent to a number of national professional accountancy organizations and regulatory bodies, universities, and individuals considered qualified to respond to the questions. Replies from sources other than recipients of the note verbale cannot be considered to be the official response of a given country, but nevertheless represent an expert opinion on the subject matter.

A total of 63 replies were received from 55 countries and from the European Union. The United Nations Conference on Trade and Development acknowledges, with appreciation, the cooperation of the respondents in providing the information requested. The response rate is a considerable improvement over last year's result on the survey of new legal requirements and/or standards governing corporate accounting at the national level.[1]

A number of respondents provided details of environmental protection laws, whereas only details of environmental accounting were requested. Clearly, environmental protection laws have an impact on the financial situation of an enterprise, and it is the recognition of environmental liabilities and provisions which is the area in which the accountancy profession has had considerable exposure to date.

Disclosure of environmental performance is another area in which the accountancy profession is involved. Previous ISAR research indicated that corporations generally are aware of the need to report information regarding their impact upon the environment. However, because many countries have not developed the required standards or even guidelines on environmental disclosures, the profession is at a great risk of compromising itself. The accountancy profession has had, in many countries, a tradition of self-regulation; hence the profession should take the initiative and formulate solid disclosure guidelines.

Without adequate guidelines, the underlying principles of accounting can be jeopardized by ad hoc and erratic disclosures. The basic accounting concepts of comparability, uniformity, reliability, to users.

A significant number of replies to the survey questionnaire drew attention to the fact that, as more and more environmental protection laws are introduced, accountants must consider the responsibilities and obligations which are placed upon corporations. Provisions are increasingly being made for environmental rehabilitation.

Only one or two replies mentioned the concept of sustainable development, which indicates that on the whole, the accountancy profession has yet to begin to develop this concept and promote its adoption. The New Zealand reply accurately pointed out that countries with a high degree of dependence upon natural resources should pay greater attention to their management. Like financial resources, environmental resources must be measured, valued and monitored by accountancy.

I. SURVEY RESULTS

A. Argentina

There are no laws or regulations in Argentina which are concerned with environmental accounting and there are no plans to formulate any. Corporations have not adopted any form of environmental accounting.

B. Australia

Currently, Australia does not have any requirements for environmental accounting, with the exception of Australian Accounting Standard No. 1022, "Accounting for the Extractive Industries". This standard requires accountants to make relevant provisions when there is an expectation that restoration costs will be incurred. The provision should be calculated as follows:

"(a) The costs of restoration work necessitated by exploration, evaluation or development activities prior to commencement of production shall be provided for at the time such activities take place and shall form part of the cost of the respective phase(s) of operations;
(b) The cost of restoration work necessitated by any mining activities after the commencement of production shall be provided for during production and shall be treated as a cost of production; and,
(c) In determining the amount to be provided in any one financial period, the balance of the provision for restoration costs, after charging

against it actual costs incurred to date, shall be reassessed in the light of expected further costs".[2]

Otherwise, environmental accounting is not regulated in Australia, so corporations can account and report as they choose. Research has shown that "out of a sample of 197 listed firms, 71 firms were identified by Deegan and Gordon as producing environmental information, either positive or negative".[3] Environmental disclosures are discretionary and information provided to the public concerning the environmental impact of the corporation is often on a subjective basis rather than being substantiated and unbiased. Bodies such as the Australian Institute of Company Directors, the Australian Institute of Management, and the Business Council of Australia have produced environmental guidelines for their members to follow. Generally, these guidelines deal with disclosure policies but also emphasize that members should be environmentally responsible.

Further, corporations which fail to operate in harmony with the environment can suffer financially through changes in consumers' spending habits and investor preferences.

C. Austria

Austria does not have any laws or regulations which specifically relate to environmental accounting. The country is in the process of entering the European Union and hence will comply with its applicable guidelines.

The Austrian Government has taken the first steps towards the development and integration of environmental performance indicators into its national accounts. This work represents a substantial contribution to the introduction of a comprehensive environmental information system which will become a part of the Austrian national statistics.

D. Bahamas

There are no laws or regulations on environmental accounting in existence or planned in the Bahamas. Ocean dumping and solid waste disposal are the major concerns of environmental reporting. It is felt that strengthened environmental accounting regulations would enhance the natural environment and the quality of life of the residents and would encourage tourism.

E. Barbados

Barbados is in the process of establishing the necessary legislative and institutional framework within which environmental accounting could be achieved. The legislation which is currently in draft format concerns environmental management and conservation but does not specifically address the issue of environmental accounting. This will be dealt with in future regulations.

F. Belgium

Currently, there are no specific Belgian laws which pertain to environmental accounting. However, the Institute of Auditors is investigating the matter and will issue a document for its members in the future. A recent survey undertaken by the Catholic University of Leuven [4] found that Belgian companies voluntarily mention environmental issues in their annual reports although this information is mainly descriptive, non-financial and partial. This finding is consistent with previous reports by ISAR [5]. Generally, only positive environmental information is disclosed.

It is felt that environmental accounting regulations need strengthening so that there is comparability of information disclosed and to ensure that environmental data reported meet quantitative and monetary standards. Furthermore, there is a possible need for specialists to assist management with environmental reporting and the principal auditor in his work. Possibly, companies with a higher environmental waste risk should publish a special report to this effect.

G. Botswana

Botswana has a number of environmental protection laws, but there are no laws which specifically pertain to environmental accounting.

H. Brazil

There are no specific provisions for environmental accounting in Brazil; however known or potential liabilities stemming from environmental damages must be disclosed or adequate provision made. The Brazilian Securities and Exchange Commission has recommended that listed companies make reference in their annual reports to relevant environmental issues. Currently, legislation is being considered which will require all corporations to undergo an environmental audit for compliance with environmental protection laws. An environmental audit is defined as the periodic and organized examination of the technical and administrative aspects related to environmental protection activities of all productive units of an entity, with the following main objectives:

(a)　To verify that all enterprises comply with the municipal, state and federal requirement to maintain and update their registrations, authorizations and licences;

(b)　To verify that all enterprises comply with restrictions and recommendations contained in the licences granted to them and in studies on environment protection relating to equipment, procedures and sites;

(c)　To verify that all enterprises comply with laws, rulings and regulations:

(i)　as to standards of emission and of environmental quality of the regions in which they are located;

(ii)　as to recovery and maintenance of related environment quality;

(d)　To evaluate the environmental protection policies of enterprise as to their:

(i)　procedures to appraise, control and prevent environmental damage in all forms;

(ii)　usage and conservation of all forms of energy;

(iii)　rational and economic use and transportation of raw materials;

(iv)　rational use, conservation, recycling and reutilization of industrial water;

(v)　minimization, recycling, treatment and disposal of solids, liquids and gaseous wastes;

(vi)　improvement of the environmental impact of production processes;

(vii) improvement of the environmental impact of products;

(viii) prevention and reduction of accidents;

(ix) training, motivation and increased awareness of personnel with respect to the care and protection of the environment; and,

(x) disclosure of enterprise policies and procedures on environmental protection and disclosure of the involuntary risks to which communities may be subject.

Some corporations disclose environmental information although at a less than optimal level. In some instances, corporations have disclosed details of anti-pollution and recovery processes in accordance with recommendations made by the Securities and Exchange Commission. However, as the requirements are not mandatory, disclosure is on an ad hoc basis and therefore more comprehensive and enforceable guidelines should be introduced.

I. Bulgaria

Environmental accounting, introduced in Bulgaria in 1992 through the Accountancy Act, is an integral part of the country's environmental management policy. It is recognized that information is required to enable correct managerial decisions to be made concerning environmental policies and in the introduction and justification of appropriate economic regulations for environmental protection.

The Accountancy Act requires completion of two types of statistical forms to produce a single source of accounting and statistical information. All companies complete a form entitled "Environmental Protection - Costs", which contains details of expenditure for environmental purposes on non-current assets such as construction, rehabilitation and preservation costs of sites, the upgrading of equipment, and intangible assets such as software and patents. The cost of maintaining these "environmentally motivated" assets is reported, in addition to the costs of environmental activities such as re-cultivation, erosion prevention, lasting or major improvements to chemicals production, biological and integrated protection, and afforestation. The costs are broken down in accordance with the essential components of the environment and its protection, namely: water; air; soils; waste management; and noise prevention. Costs are also classified by activities, such as forestry and agricultural, for monitoring the development and management of wildlife and fisheries. The environmental sites at which costs have been incurred must also be reported.

Currently, the Government is formulating a standard which will deal with the environmental costs incurred in preserving the ecological balance and eliminating damage to the environment. Corporations will be required to report, in an appendix to their annual accounting report, information on environmental protection costs such as the fees paid for the right to pollute within admissible levels and the contingency costs of environmental degradation. These costs are to be reported for the current year and estimated for the following year. Violations of the regulations or incorrect information previously provided must be reported by corporations.

In spite of these accomplishments it is still felt that regulations to control the appropriateness and scope of environmental accounting information require strengthening.

J. Canada

Laws and regulations exist concerning the disclosure of quantitative and/or qualitative environmental information. The following areas in particular are dealt with in the existing statutes and standards:

(a) "When reasonably determinable, provisions should be made for future removal and site restoration costs, net of expected recoveries, in a rational and systematic manner by charges to income".[6] Such provisions are made either as a result of environmental law or contracts, or because the enterprise has established a policy to restore a site;

(b) Contingent liabilities, whilst non-specific, are also applicable to environmental considerations;

(c) Securities commissions at the provincial level require disclosure of "financial or operational effects of environmental protection requirements on the capital expenditures, earnings and competitive position of the issuer for the current fiscal year and any expected impact on future years".[7]

There are currently no laws or proposals on the broader environmental accounting elements such as sustainable development, full environmental cost accounting and environmental performance indicators.

The revision of existing accounting standards which concern the way in which environmental costs and liabilities are treated within the existing financial reporting model is under consideration by the Canadian Institute of Chartered Accountants (CICA). In October 1994, CICA, in association with other groups in Canada, published a study entitled Reporting on Environmental Performance, which provides guidance to organizations on the various matters that should be taken into consideration when reporting on environmental performance. As stated in the foreword to the publication, "One of the objectives was to develop a framework for reporting in accordance with standards and expectations relating to environmental conduct and responsibility."

Interpretation of the accountancy guidelines as they apply to environmental costs and liabilities results in variations in the treatment of pollution control and site clean-up costs and there are difficulties with the recognition and measurement of environmental liabilities, and with the disclosure of environmental expenditures. A CICA 1993 publication entitled Environmental Costs and Liabilities: Accounting and Financial Reporting Issues gives recommendations for appropriate practice.

CICA feels that environmental performance, together with the use of indicators and additional environmental information (quantitative and qualitative) that are currently outside the scope of existing financial reporting, is an area of priority for the development of environmental accounting.

Furthermore, it is being suggested that the investment community could use information concerning regulatory compliance in measuring earnings, asset values, liabilities, risks and equity. Greater disclosure could lead to increased awareness of the environmental impact and performance of corporations and more attention might be spent on these matters by directors and management.

K. China

Although China has a number of environmental protection laws, there are no laws which specifically relate to environmental accounting. Disclosure of the following is considered to be important:

(a)   The costs incurred by corporations to reduce and prevent pollution and restore the environment;

(b)   Pollution-related costs and losses, such as compensation for environmental damage, fines for the breach of environmental protection laws, payments to the State for pollution licences, and clean-up costs;

(c)   The social cost of pollution.

L. Colombia

Colombian corporations are required to disclose quantitative and qualitative environmental information in an annual report to the Government. This report contains information on the following: the extent to which the corporation conforms with environmental policies; the degree of environmental degradation caused by commercial activities; and the status of the management and conservation of natural resources. Furthermore, corporations whose operations have an impact on the environment must undergo environmental audits. For this reason environmental indicators are currently being devised. Also corporations which propose investment projects must provide an estimate of the expected environmental impact.

There are special measures on environmental protection and rehabilitation costs which break the data down by areas of impact such as water and air pollution, solid and liquid wastes, and noise. The information is divided into costs incurred in present and past operations. Details of expenditure on technological changes and innovations, research and development, and environmental impact studies must also be disclosed.

M. Croatia

Croatia is currently developing a number of environmental laws. Until such time as national standards have been developed, those of the European Union prevail.

N. Cyprus

There are no specific laws and regulations governing environmental accounting in Cyprus. The disclosure of environmental information occurs on a voluntary basis and generally in conformity with relevant international accounting standards. A comprehensive environmental framework law is under consideration. The following areas are considered to be a priority for environmental accounting:

(a)   Environmental revenues and capital expenditures;
(b)   Fines imposed on enterprises and corrective measures taken;
(c)   Contingent liabilities arising from failure to adopt environmental provisions.

O. Czech Republic

There are no laws in the Czech Republic which specifically deal with environmental accounting but a bill on the subject is under consideration. Tax laws allow a deduction for the creation of reserves to repair environmental damage. It is considered that the development of systems and procedures to include in the costs of production and sales the total costs related to depletion and restoration of the environment is the highest priority for new environmental accounting regulations. The Government is interested in discussing priority problems with representatives of other countries that are leaders in the field of environmental accounting.

P. Denmark

Denmark has incorporated the European Union's eco-management and audit scheme (see the

section on the European Union below) into its own laws, and has established an environmental control agency. It is considered important that non-industrial sectors should be involved in this scheme, as well as small and medium-sized corporations.

Moreover, the Danish environmental legislation requires enterprises whose activities cause major environmental pollution to seek environmental approval from the authorities.

Q. Ecuador

Ecuador does not have environmental accounting laws; however the accountancy profession recognizes the need for such laws and has made some preliminary studies of current international regulations to identify suitable standards for Ecuador. Current environmental accounting practices are ad hoc and depend on the policies of individual corporations.

R. Egypt

Laws require corporations to keep a register of their impact upon the environment; however, this information is not always disclosed in corporate annual accounts. More comprehensive environmental protection and preservation laws that are now being introduced are likely to lead to new environmental accounting regulations.

S. Fiji

At present Fiji has no environmental accounting laws or regulations, and does not plan to formulate any.

T. Finland

Finnish corporations have provided environmental information for a number of years, primarily on a voluntary basis but also in response to legislative requirements. Furthermore, public involvement in projects with significant environmental effects is increasing. Companies are obliged to report publicly on their emissions, chemical and energy usage and accidents, among other environmental matters. Finland will adopt the European Union regulations on voluntary participation in the environmental management and audit scheme.

Corporations provide environmental information in special reports or within the annual report. In some instances third parties have verified the environmental information contained in such reports. There is a trend to include environmental information in annual reports because it is felt that environmental protection and management should be an integral part of industrial activities and hence should be reported as such rather than in the isolation of a special purpose report. Certain industries, such as the forestry and chemical industries, publish sectoral environmental reports.

It is suggested that pressure from clients, consumers and competitors can more effectively lead to improved environmental performance and management practices than strict mandatory requirements. International guidelines can also be effective in this area.

U. France

France has developed an ecological balance sheet which concerns the relationship between an enterprise and the environment. The depletion of natural resources and the problems of pollution are linked to the production process. Furthermore, corporations are increasingly aware of the benefits of integrating environmental protection into company management policies. A social balance sheet has also been formulated which pertains to the enterprise and its staff.

Both of these "new" balance sheets are in the format of a table of environmental issues and are expressed in monetary terms. Information is provided on the acquisition and utilization of equipment used to reduce pollution, recycled by-products, and reduced energy and raw materials consumption. This information may be assimilated into one table and may include expenditures and benefits on a number of items such as: environmental protection, royalties, licences and compensation costs for the prevention of pollution; operation and maintenance of equipment specifically purchased to help preserve the environment; grants or loans obtained at preferential rates to assist in the purchase of specific equipment.

An alternative proposal is to use an accounting system which highlights the company's contribution to environmental management. This system would involve three tables showing:

(a) An analysis of the productive activity (conversion of natural resources into products, sub-products and residues);

(b) The means employed and the results achieved in environmental protection;

(c) The impact of the company's activities on the national heritage of natural and man-made resources.

The aim of this "ecological balance sheet" is to help determine the cost of combating pollution and contribute to the rational use of energy and raw materials.

Currently, there are no binding provisions with regard to the inclusion of environmental information in annual reports; corporations may adopt procedures which differ from the norm. Some companies may have a section of their annual report devoted to environmental policies such as: the objectives pursued in processing waste and combating environmental pollution; the development of technology and new products which reduce energy consumption and pollution; and the management of major environmental hazards. Financial information is also provided, such as reserves for clean-up costs.

It is believed that there is a need for greater disclosure of environmental protection information and more guidance on contents and formats in order to facilitate better comparisons between various companies. Furthermore, information should be broken down into major themes, with quantitative data and uniform ratios (environmental performance indicators) to enable users to perceive with greater precision the efforts made by enterprises in the field of environmental management and risks.

V. Gabon

The government has recently adopted a law concerning the protection and improvement of the environment however, currently there are no environmental accounting laws or regulations.

W. Georgia

Georgia has no regulations for, or voluntary practices on, environmental accounting. Natural resources and environmental protection are considered to be priority areas to be addressed within corporate accounting and national accounts.

X. Germany

Germany has no accounting requirements specifically relating to the environment, with the exception of:

(a)   An obligation to report investment activities in the field of waste disposal, noise reduction, and decreases in air and water pollution; and,

(b)   The responsibility to record contingent liabilities for potential losses which may arise as a result of environmental protection laws. These laws are quite comprehensive and place considerable demands upon corporations. For example, manufacturers are obliged to dispose of consumers' wastes at the end of a product's life.

The concept of depreciation has been expanded to include environmental contamination and obsolescence of equipment due to changes in environmental protection laws. The tax deductibility of environmental cost provisions is being considered by the Federal Ministry of Finance.

On a voluntary basis some corporations provide environmental information in the form of an input/output analysis for products and processes and details of their environmental policies and measures taken to protect the environment. More comprehensive environmental disclosures are made on a non-compulsory basis by large corporations. The relevant European Commission regulations are possibly a factor in this. Corporations are also

under an obligation to provide information which is of significance to financial markets. A number of large corporations have developed environmental balance sheets and have also undergone environmental audits.

Priority areas for environmental accounting disclosures include: details of expenses incurred to attain environmental goals, such as new technology to reduce air pollutants, and other wastes from production processes; emission fees; and provisions for environmental risks. It is felt that environmental accounting regulations concerning provisions against environmental protection and clean-up costs need strengthening. If corporations do not make adequate allowances, the State may have to assume financial responsibility.

Y. Hungary

Although there are no comprehensive environmental accounting laws, certain environmental protection disclosures must be made. For example, annual company reports must contain information on: tangible assets acquired to protect the environment; the quantity and quality of the year's opening and closing volumes and values of hazardous waste (which is also classified by rating standards defined by law); and details of provisions made for environmental responsibilities and liabilities and for any contingencies.

Special tax concessions and depreciation allowances are available to encourage investments in equipment based on environmentally sound technology. Licences are required and fees are charged for conducting activities which are detrimental to the environment. Rehabilitation and clean-up costs must be considered as priority claims during insolvency procedures.

Legislation is being drafted to provide greater environmental protection. These laws do not specifically refer to environmental accounting, but by implication will require the expansion of existing accounting requirements, particularly concerning provisions for environmental protection and rehabilitation. Revised environmental protection laws could expand the role of accountancy in the cost/benefit analysis of management decisions.

Environmental awareness has affected the privatization process because the cost of previous environmental degradation and the use of unsound environmental technologies must be taken into account during the valuation process. Asset values are affected and there is a need to make provision for the rehabilitation of past and present environmental damage.

Z. India

An environmental statement concerning water, raw materials, pollution, solid and hazardous wastes and disposal practices is required of all industrial organizations. Furthermore, company directors in their annual report must disclose measures taken for environmental protection. India does not require environmental accounting beyond these disclosure requirements. It is considered desirable for the impact of environmental protection measures on the profitability of an entity to be disclosed in annual reports.

AA. Ireland

Ireland has no laws or regulations in respect of environmental accounting. Future laws in this regard will be those prescribed by the European Union. Reporting of environmental matters to date has been on an ad hoc basis, with environmental policies and internal structures established to comply with environmental regulations receiving particular attention.

It is felt that priority should be given to:

(a)   The provision of environmental restoration and clean-up costs;
(b)   A statement to outline expenditures incurred in implementing plans to achieve environmental objectives;
(c)   The assessment and disclosure in a separate statement of the costs and benefits of a corporation's environment-related activities.

Furthermore, it is recommended that compliance with environmental accounting standards should be strictly enforced and breaches thereof should be penalized.

BB. Italy

Italy follows the European Union requirements for environmental accounting. It is believed that only environmental issues which have a direct impact upon operating performance or financial position should be disclosed. This disclosure should be made in:

(a)   The balance sheet, by way of disclosing details of provisions;
(b)   The notes, in particular by disclosing details of valuation methods used for accounts, such as extraordinary items, contingent environmental liabilities, and environmental expenditures capitalized and/or charged to the profit and loss account;
(c) Other sections of the annual report, with a description of environmental issues and the management's response, environmental policies and implementation of environmental protection measures, details of government incentives relating to environmental protection measures, and compliance with existing and expected legislation concerning environmental protection.

There should be a reference to a separate environmental report when one exists.

It is felt that a great deal of the environmental information available at present is of a qualitative nature without an assessment of the financial implications. More quantitative information could increase the value of information to users. The introduction of full-cost or sustainable development accounting is still in its infancy, with many issues still to be resolved.

CC. Japan

There are no environmental accounting laws in Japan nor is their formulation foreseen. However, if environmental issues have an impact upon a company's financial position and operating results for the current or future periods, the impact is disclosed in either the annual report or financial statements. A Japanese environment agency conducted a survey which revealed that 29.7 per cent of corporations report separately environmental costs and investments, in particular those dealing with pollution prevention and the cost of waste disposal. The survey also revealed that 36.7 per cent of corporations responded positively to the idea of formulating environmental disclosure guidelines.

DD. Kenya

Kenya does not have environmental accounting laws and regulations, nor are any being planned. International standards, if appropriate, might be implemented in the future.

EE. Lesotho

Lesotho does not have environmental accounting laws or regulations and none are currently being formulated. It is felt that priority should be given to disclosing in financial statements the costs incurred in compensating for the ecological damage caused by significant development projects in the country.

FF. Luxembourg

There are no environmental accounting laws or regulations in Luxembourg.

GG. Malaysia

Although environmental protection laws exist in Malaysia, there are currently no laws, nor are any being planned, which pertain to environmental accountancy. It is considered that details of capital expenditures and operating expenses incurred specifically to improve pollution control, occupational health and safety, product safety and waste management are areas of priority in the development of environmental accounting standards.

HH. Mexico

At present Mexico does not have, nor does it plan to develop, environmental accounting laws and regulations. Businesses are able to depreciate equipment used to prevent and control pollution at accelerated rates for taxation purposes.

II. Morocco

Currently there are no environmental accounting laws or regulations in place in Morocco, nor are any being considered. Furthermore, there are no known environmental accounting practices followed by corporations. It is felt that priority should be given to the development of appropriate standards for supplementary information in annual financial statements to disclose:

(a) Qualitative information concerning environmental damage caused by corporations; and,

(b) Quantitative information on investment and development expenditures associated with environmental protection.

Improved environmental accounting could lead to greater sensitivity to environmental problems by management and heightened public awareness and appreciation of a corporation's environmental performance.

JJ. Nepal

Although Nepal has a number of environmental protection laws, there are no laws or regulations which specifically relate to environmental accounting.

KK. Netherlands

The Netherlands has adopted an Environmental Protection Act which allows government authorities to monitor the environmental impact of corporations. Environmental accounting is not specifically regulated, although there is currently a bill on environmental reporting which requires corporations that place a large or moderate burden on the environment, or which pose special environmental risks, to report to the public on their environmental impact and their efforts to reduce damages.

The proposed annual environmental report is expected to contain legally prescribed requirements and require submission of completed documents to appropriate government authorities. The report will likely contain information of a quantitative and qualitative nature concerning: the emission of particular substances and noises; energy and raw material consumption; and the generation of waste and its treatment. Details of the measures taken and costs involved in reducing the extent of such emission, consumption or waste may also be included. Data may be required on the environmental impact of a corporation's products, in the production and waste disposal phases for example, and materials recycling and site restoration possibilities. Possible environmental indices could be used as a medium to report the complete production, utilization and disposal cycles of the industrial processes. Furthermore, the report could also cover wider areas such as "transport flows to and from the corporation, research on the protection and rehabilitation of the environment, internal and external audits, public information campaigns and training given to employees".[8]

It is suggested that future trends should also be indicated in the data included in the report. The public would then be in a better position to gauge the extent to which the environmental impact of the corporation will increase or decrease. Difficulties in forecasting may restrict the forecast period to one year.

The bill does not aim to establish inflexible rigorous requirements. It is hoped that corporations will continue to develop the environmental reports that they already prepare on a voluntary basis; nevertheless the proposed special report must at least meet the minimum requirements that will be established.

Research in the country has shown that "of the 128 most polluting enterprises, some 30 have published or plan to publish environmental reports on a voluntary basis. This number is higher if enterprises are included which incorporate a description of their environmental policies in their corporate annual accounts or annual report".[9]

LL. New Zealand

Although New Zealand does not have and is not preparing any laws specifically related to environmental accounting, there is a recent law which deals with environmental management (Resource Management Act, 1991). The law controls the use of natural resources and has as its stated purpose "...to promote the sustainable management of natural and physical resources". It defines "sustainable management" as:

"...managing the use, development and protection of natural and physical resources in a way, or at a rate, which enable peoples and communities to provide for their social, economic and cultural well-being while:

(a)   Sustaining the potential of natural and physical resources (excluding minerals) to meet the reasonably foreseeable needs of future generations;

(b)   Safeguarding the life-supporting capacity of air, water, soil and ecosystems; and

(c)   Avoiding, remedying or mitigating any adverse effects of activities on the environment".

The law prescribes substantial fines and even imprisonment for infringements. Furthermore, the present owner or occupier of land is held responsible for contamination of that land, regardless of when the damage actually occurred.

New Zealand's accounting standards do not differentiate between environmental and other liabilities. Consequently, all liabilities must be provided for in a corporation's financial statements as either an actual liability, a provision, or as a contingency disclosed in the notes.

A large number of corporations are working towards improved environmental performance. A 1992 country survey revealed that 36 per cent of the largest corporations described their environmental activities and policies in their annual report. Although no corporations actually prepare "environmental accounts", some use non-financial measures to convey environmental results. A forest and timber products company has announced that it will publish a "greenhouse gas balance sheet", which will show to what extent greenhouse gas emissions from its manufacturing processes are offset by the absorption of the gases in its forestry plantation.

The New Zealand accountancy profession has recognized the potential business development opportunities in the field of environmental audits. Furthermore, to better assist their clients to gain the optimum advantages available within a system of environmental management, accountants must be completely familiar with environmental management laws. Accountants in some respects face a dilemma in dealing with provisions for yet-to-be- detected contamination which, if recognized within financial statements, is effectively an admission of responsibility.

Current accounting standards do not measure the productive capacity of a corporation's natural resources. Consequently, the long-term income-generating capacity of assets (and therefore their value) may be understated. This is particularly important to an economy such as New Zealand's whose natural asset base is critical to a relatively high proportion of its gross national product.

MM. Pakistan

The concept of environmental accounting is new in Pakistan, and laws and regulations have not yet been introduced in this regard.

NN. Panama

Currently there are no environmental accounting laws or regulations in Panama. It is suggested that priority areas for the formulation of such regulations are the disclosure of deforestation, soil degradation and pollution. Recently, a commission was established to develop an appropriate set of environmental statistics which will be used in national accounting with the aim of valuing natural resources.

OO. Philippines

There are no rules or regulations regarding environmental accounting in the Philippines, although there are a number of laws which deal with environmental protection.

PP. Poland

There are many laws and regulations in Poland with respect to the environment. One law specifies provisions for environmental damage caused by mining operations. However, there are no laws which are concerned with environmental accounting. The environmental protection laws give rise to responsibilities and potential liabilities for which possible obligations may need to be recorded. Fiscal instruments are in place for environmental matters, such as taxation incentives for positive activities and penalties and fines for environmental damage.

Poland is currently formulating a new accounting act. Although environmental accounting is not explicitly included, it will appear in subsequent revisions to the act. It is suggested that the valuation and recognition of specific environmental expenses or losses and the disclosure of environmental information in financial statements are areas in which greater guidance should be offered.

Corporations whose activities present a significant threat to the environment, such as those in the chemical industry, are inclined to provide more comprehensive disclosures. Details reported include:

(a) Environmental expenses and penalties;
(b) Environmentally sound investments;
(c) Future and current environmental policies; and,
(d) Details of the manner of cooperation between the corporation and the government's environmental agency.

The above information is usually provided for the current year and preceding three years. Provisions are sometimes made for obligations which arise due to the environmental protection laws, but the provisions are not allowable in the calculation of taxable income until the expenditures are actually made.

It is reported that priority should be given to the establishment of environmental accounting standards for the disclosure of:

(a) The formal policies and programme on protection of the environment;
(b) Improvements in environmental performance;
(c) Mitigation of environmental damage;

(d)   The policy used by corporations to recognize environmental liabilities and provisions for contingent liabilities;

(e)   Investments motivated by environmental considerations and the financial impact of such considerations.

It is noted that although environmental management has been the subject of much debate the accountancy profession and accountancy laws do not yet provide a sufficient answer to the problem.

QQ. Korea

There are no environmental accounting laws nor are any foreseen in the Republic of Korea. The following areas of environmental accounting are considered a priority:

(a)   Compilation and integration of environmental and economic data;

(b)   Institutional arrangements across the various environment- related government agencies for the development and promotion of an environmental accounting system;

(c)   Development of a system to monitor the quality of the environment and the natural resources base;

(d)   Linkage between physical information and economic data by valuation methods.

The framework must be appropriate on a country-by-country basis.

It is suggested that difficulties in the interpretation and usage of the existing physical information within national accounts require the development of a scientific method to incorporate the data, especially regarding the quality of the environment.

RR. Russian Federation

Environmental accounting in the Russian Federation is limited to the statistical reporting of waste quantities and of current year expenditures on environmental protection measures and management. No changes to accounting laws and regulations are envisaged in the immediate future to incorporate environmental considerations.

SS. Singapore

Singapore does not have laws and regulations relating to environmental accounting, nor or any being formulated. Corporations do not depart from established legal requirements. However, it is suggested that the disclosure of contingent liabilities and the social implications of environmental activities are an area of priority for which specific standards should be developed.

TT. South Africa

There are no specific environmental accounting laws or regulations in South Africa. However, there are a number of laws which relate to environmental protection and management.

Recently, the University of Pretoria, in cooperation with the Chartered Association of Certified Accountants (a professional body based in the United Kingdom), studied environmental reporting in South Africa. This study found that "...only 17 per cent of the top listed companies in South Africa mention an objective or mission regarding the environment in their annual reports...only 7 per cent of the top listed companies in South Africa indicate that they have met their objectives regarding the environment."[10]

Generally, environmental reporting is restricted to the positive aspects of performance, and none of the company reports mentioned any negative aspects. Only 8 per cent of the companies surveyed mentioned an environmental audit, of which only 15 per cent are independently attested. Generally, disclosure of environmental matters is done on an ad hoc basis.[11]

South Africa will use the results of this survey to develop appropriate standards for environmental reporting which will be broadly based on the Institute of Chartered Accountants in England and Wales framework.[12] (See the section on the United Kingdom of Great Britain and Northern Ireland below.)

UU. Spain

Although there are no specific environmental accounting provisions within the present laws and regulations, the Companies Act and the General Accounting Plan stipulate that any kind of risk must be accounted for; therefore environmental risks must be accounted for. Currently, the Chart of Accounts for the water distribution industry is being elaborated. According to the current state of the works, corporations in the water distribution sector will be required to provide information on the ecological impact of their activities. However there are no general rules being formulated for environmental accounting on a broader basis. A number of large corporations have disclosed environmental information on a voluntary basis, but as a general rule spanish corporations do not provide detailed environmental information, and the details that are provided are of a qualitative rather than a quantitative nature.

It is perceived that enterprises have difficulties in providing environmental information related to the environmental costs that are not obligatory for enterprises in order to comply with environmental protection laws. It is suggested that environmental information other than that which has a direct relationship to property values, the financial situation, or the operating results of a corporation should be disclosed in a special purpose report. If financial statements are required to provide more comprehensive environmental information, management may as a result adopt more environmentally sound policies. Consumer pressure can also contribute to this end.

VV. Sudan

Environmental legislation exists to protect many aspects of the environment, although none relate specifically to environmental accounting. Areas which need development are: disclosures on the environmental impact of development projects; national accounts records of natural resources; and an appropriate methodology in the current accounting system to account for natural resources.

WW. Switzerland

Swiss accountancy laws and regulations do not require corporations to provide details specifically on environment-related issues, except to the extent that environmental considerations have an influence on the financial position of the corporation as a consequence of complying with environmental protection laws. Nevertheless, a number of corporations voluntarily provide environmental information, the details of which vary significantly.

No new laws concerning the environment and accounting are under development at present. Priority areas of environmental accounting include: pollution, in particular from emissions; investments for environment protection; and the extent to which there could be potential future financial impacts due to environmental risks.

XX. Thailand

Although there are laws which relate to the enhancement and conservation of the environment in Thailand, there are currently no environmental accounting standards. Some companies voluntarily disclose environmental concerns in their annual reports. Priority aspects of environmental disclosure include:

(a) Accounting policies and an environmental management plan, and the corporation's remedial activities and concept of sustainable development;

(b) Legal proceedings and compliance with legislative requirements;

(c) Research and development and capital expenditure related to the environment;

(d) Environment-related expenditures, liabilities and provisions;

(e) Any funds received from the public or State to support projects or activities specifically undertaken to improve the quality of the environment as a result of commercial activities.

A number of environmental accounting regulations should be developed, concerning: environmental audit expenses; research and development expenditures and the measurement and allocation of production costs; environmentally motivated investments; and environmental provisions and liabilities.

YY. The former Yugoslav Republic of Macedonia

The country has several laws which deal with environmental protection, although there are none which specifically relate to environmental accounting. The main sections of the laws are on the atmosphere, water, natural resources, forests, land and noise. All corporations are obliged to measure and maintain records of pollution that results from their production processes. New environmental protection acts are expected to be based upon European Union recommendations and other international standards.

ZZ. Trinidad and Tobago

At present there are no environmental accounting standards or guidelines in Trinidad and Tobago, nor are any being proposed.

AAA. Turkey

Turkey does not have any laws or regulations which concern environmental accounting.

BBB. United Kingdom and Great Britain and Northern Ireland

Currently, there are no environmental accounting regulations or laws in the United Kingdom, not are any being formulated. However, a number of corporations, particularly the largest ones, have introduced environmental disclosure on a voluntary basis. Furthermore, the recognition of environmental liabilities is growing, in addition to the disclosure of environmental provisions.

It is recognized that previous ISAR guidelines on environmental disclosures are widely accepted and could be elaborated. Environmental accounting on the whole needs strengthening for the benefit of society, to promote openness within corporations, and to foster greater self-esteem within the company itself.

Pressure from the public, both from individuals and corporations, can have an impact upon the policies adopted by corporations. For instance, a British hardware chain recently requested each of its suppliers to provide environmental audits for each of the products it purchased, otherwise the product might be excluded from the retailer's buying list. Furthermore, the Chartered Association of Certified Accountants has an annual competition to identify the corporation with the most comprehensive environmental accounting in its annual report.

A recent study by The Institute of Chartered Accountants in England and Wales [13] recommended a framework for environmental reporting which includes the following elements:

(a) Details of the corporation's environmental policy, not in the sense of an immediately attainable goal but more as an ideal to which the corporation will continuously strive. Also a mission statement should be prepared which describes the day-to-day efforts of the corporation to control its environmental impact;

(b) A statement of concrete objectives, preferably stated in specific, auditable terms, that are attainable within a relatively short period of time;

(c) Narrative disclosure of the company's activities which address core business issues rather than peripheral matters. These activities should be disclosed comprehensively on a site-by-site basis, be comparable to some acceptable predetermined standards, and be formulated in specific rather than general terms;

(d) Quantitative disclosure of the corporation's environmental impact. The data should be both technical and financial, preferably integrated where possible. The following topics should be included: emission levels; energy consumption; noise levels; waste production and recycling;

(e) Quantitative disclosure of the following financial information to the extent that it is related to the environment:

(i) Policy disclosure of the accounting treatment and the environment-related aspects of financial statements;

(ii) Contingent liabilities;

(iii) Fines and penalties levied under environment protection laws;

(iv) Rehabilitation costs of hazardous waste disposal sites;

(v)   Costs incurred to dispose of wastes;

(vi)   Future costs which may arise from closure or abandonment of operating facilities;

(vii)   Costs associated with the acquisition, installation and operation of environmentally sound equipment;

(viii)   Costs associated with the redesign of products and processes to reduce reliance on hazardous materials and to minimize waste;

(ix)   Legal, consultancy and administrative costs incurred to comply with environmental laws and regulations.

The study suggested that there is a need for further research into the following:

(a)   The possibility of extending ordinary accounting conventions to encompass the external costs and benefits of a company's environmental impact;

(b)   Understanding the use made of environmental information by participants in the capital markets;

(c)   The development of reliable measures of environmental performance and their impact;

(d)   The extent to which management information systems are being adapted to capture, process and report environmental data;

(e)   The likely impacts of fiscal incentives on corporate behaviour.

CCC. United States of America

Although United States accounting standards do not comprehensively address environmental accounting issues, there are a number of rules on the reporting and recording of costs and liabilities that pertain to environmental matters. These are:

(a)   A Financial Accounting Standards Board (FASB) Standard on "Accounting for Contingencies"[14], and a Standards Interpretation, "Reasonable Estimation of the Amount of a Loss"[15] which provide guidance on the recognition and disclosure of contingent liabilities, including environmental liabilities;

(b)   An Emerging Issues Task Force (EITF) report, "Accounting for the Costs of Asbestos Removal"[16], which considers whether

costs should be capitalized or treated as an expense;

(c)   An EITF report, "Capitalization of Costs to Treat Environmental Contamination"[17], which also addresses the issue of capitalization or expensing of expenditures;

(d)   An EITF report, "Accounting for Environmental Liabilities"[18], which addresses the measurement of a loss when recoveries are possible and whether the liability should be measured on a discounted or un-discounted basis.

The Securities and Exchange Commission has issued guidelines that pertain to the recognition, measurement and disclosure of environmental liabilities. The aim of these guidelines is to ensure that users of financial statements have sufficient pertinent information with which to make well-informed decisions.

Otherwise, corporations are not required to provide any other environmental information. Nevertheless, a number of corporations have begun to produce separate reports on the environmental impact of their operations.

The Environmental Protection Agency (EPA) requires companies to file a variety of reports concerning past, present and potential future pollution and contamination. The Agency has undertaken a programme to monitor hazardous waste materials as they move from "cradle to grave". The reports are available to the public.

The American Institute of Certified Public Accountants is drafting an accounting guide that will address accounting for environmental liabilities associated with past contamination. The guide will not provide guidance for ongoing or future contamination. The FASB will also consider accounting for nuclear decommissioning costs, among its other agenda items.

It is suggested that current accounting guidelines do not adequately address a corporation's future obligations to restore natural resources extraction sites. However, industries such as mining, electric utilities and oil and gas extraction are required to restore their extraction sites upon the termination of operations.

A survey of corporate environmental accounting was undertaken by Price Waterhouse in 1992. The following are the most salient findings[19]:

(a)   One third of the respondent companies now have written environmental accounting policies...26 per cent of respondents disclose their environmental accounting policies in the notes to the financial statements;

(b)   Known exposures to environmental clean-up liabilities have not been recorded by 62 per cent of the respondent corporations. The measurement and determination of the timing of these liabilities is a difficult task;

(c)   Many corporations are utilizing their internal audit staffs to review compliance with government statutes, environmental accounting procedures, and adherence to internal policies;

(d)   Typically, only certain extractive industries and the nuclear power industry accrue for future site restoration obligations.

DDD. Uruguay

Uruguay does not have laws and regulations specifically relating to environmental accounting, nor are any being drafted. Corporations generally do not adopt environmental accounting on a voluntary basis.

EEE. European Union

The European Commission has prepared an Eco-Management and Auditing Scheme, which is a series of guidelines for voluntary participation by industrial corporations. "The objective of this scheme is to promote continuous improvements in the environmental performance of industrial activities by:

(a)   The establishment and implementation of environmental polices, programmes and management systems by companies in relation to their operating sites;

(b)   The systematic, objective and periodic evaluation of the performance of the above activities;

(c)   The provision of information on environmental performance to the public."[20]

II. CONCLUSIONS AND RECOMMENDATIONS

The questionnaire that was circulated requested information as to which areas of environmental accounting required strengthening in each country. Many respondents replied solely in terms of:

(a)   Environmental disclosures on either a quantitative or qualitative basis; and/or,

(b)   Environmental provisions and liabilities.

However, the questionnaire defined environmental accounting in the broader terms of: (i) disclosure; (ii) accounting for sustainable development or full cost accounting where industry practice has been modified to ensure the mutual long-term viability of the commercial activity and the environment; and, (iii) environmental performance indicators which may or may not link environmental information directly to financial data of the corporation.

The replies concerning recommendations for strengthening environmental disclosure were far more numerous and comprehensive than those concerning any other area of environmental accounting as defined by the questionnaire. It remains unclear whether the general perception is that environmental accounting should not extend beyond the two above-mentioned areas or that significant efforts need to be made by the profession to deal with environmental accounting in much broader terms.

The areas which ISAR has suggested previously for environmental expenditure disclosures[21] are supported by the questionnaire replies. From the replies it is not completely clear whether reporting should include quantitative and qualitative details on wastes, recycling and energy consumption. Areas of disclosure suggested in the replies, either from a monetary or physical viewpoint, for which ISAR has previously not provided guidance for disclosure, include: research and development; noise and vibration; fines and penalties; protection of land; and the social impact of industrial processes.

Also, environmental liabilities and provisions are areas which were suggested as requiring greater development in terms of accounting guidelines. Furthermore, the "non-traditional" areas of environmental accounting which the respondents suggested as areas for development include:

(a)   Full-cost environmental accounting;

(b)   Environmental performance indicators;

(c)   The applicability of the traditional historical cost accounting model to the more demanding and challenging concept of accounting for sustainable development;

(d)   Environmental auditing;

(e)   The relationship between corporate environmental accounting and national accounting for natural resources.

A number of replies indicated that concern by the general public and pressure from environmentalists were two factors responsible for corporations adopting environmental accounting practices on a voluntary basis. However, this approach can lead to inconsistent disclosures, which

undermines the credibility of accounting and reporting.

According to an article which appeared in the 16 November 1994 edition of the Financial Times, financial analysts "... find it difficult to identify potential environmental risks because of the poor quality of information made available by companies...banks are paying more attention to their exposure to environmental risks and investors are starting to show concern about risks posed by companies with poor environmental records... Most of the information [environmental], including specific quantitative details, is arbitrary and self-determined. In very few cases is it possible to draw direct conclusions as to the implications for revenues and profit margins."

In many countries the accountancy profession is self-regulatory. In order to overcome some of the above-mentioned deficiencies, a more vigorous approach may need to be taken so as to maintain the integrity of the accountancy profession and to best serve the government, labour and other users of corporate reporting.

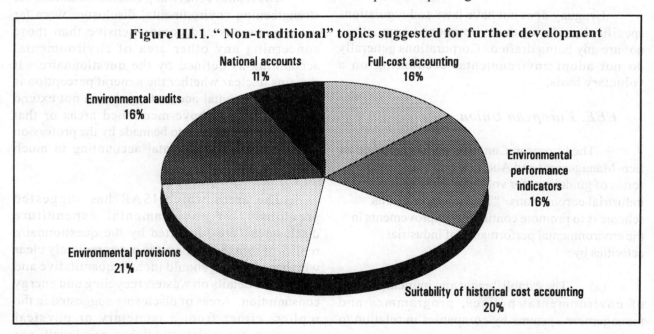

Figure III.1. " Non-traditional" topics suggested for further development

Notes

1. United Nations, E/C.10/AC.3/1994/3.
2. Australian Accounting Standards Board, AASB 1022 "Accounting for the Extractive Industries".
3. Institute of Chartered Accountants in Australia, "The Environment: an Accountant's Dilemma", Charter, August 1994, pp. 64-66.
4. Van Nuffel, L., Lin, L. and Lefebvre, C., "The External Corporate Environmental Reporting".
5. United Nations, E/C.10/AC.3/1992/3 and E/C.10/AC.3/1994/4.
6. Canadian Institute of Chartered Accountants, CICA Handbook, Section 3060.39, "Capital Assets".
7. Ontario Securities Commission, Policy Statement No.5.10, "Annual Information Form and Management Discussion and Analysis of Financial Condition and Results of Operations - Policies".
8. Extension of the Environmental Protection Act with regard to Environmental Reporting, bill and main passages of the explanatory memorandum, p. 17, draft of 24 January 1994, Ministry of Housing, Spatial Planning and the Environment, the Netherlands.
9. Environmental Reporting: Status Report for the Netherlands, KPMG Milieu, September 1993.
10. Green Reporting in the Republic of South Africa, first edition. School of Accounting Sciences, University of Pretoria, p. 53.
11. Ibid.
12. Business, Accountancy and the Environment: A Policy and Research Agenda, 1992, Institute of Chartered Accountants in England and Wales.
13. Business, Accountancy and the Environment: A Policy and Research Agenda, Institute of Chartered Accountants in England and Wales, 1992.
14. Financial Accounting Standards Board, Statement of Financial Accounting Standards No. 5.
15. Financial Accounting Standards Board, Interpretation No. 14.
16. Emerging Issues Task Force, issue No. 89-13.
17. Emerging Issues Task Force, issue No.90-8.
18. Emerging Issues Task Force, issue No.93-5.
19. Accounting for Environmental Compliance: Crossroads of GAAP, Engineering, and Goverment, Price Waterhouse, Second in a series, 1992, pp. 2-3.
20. European Council, Regulation No. 1836/93.
21. Conclusions on Accounting and Reporting by Transnational Corporations, Intergovernmental Working Group of Experts on International Standards of Accounting and Reporting, 1994, pp. 31-32.

Chapter IV

DISCLOSURE BY TRANSNATIONAL CORPORATIONS OF ENVIRONMENTAL MATTERS AT THE NATIONAL LEVEL

Report by the UNCTAD secretariat

Summary

Many transnational corporations (TNCs) adopt differing voluntary environmental accounting disclosure standards in their home and host countries. This report is a comparison of environmental accounting disclosures by TNCs in India, Malaysia and the Philippines as host countries of TNCs vis-à-vis the reporting by the same transnational corporations in their home countries. TNCs in the food and chemical industries were selected for study. It was found that there are considerable differences in reporting by TNCs at the local level and home country level.

I. INTRODUCTION

This report focuses on environmental disclosures by transnational corporations. It is restricted to a comparison between the environmental reporting practices by transnational corporations in host countries (all of which are developing countries in this project) with their home countries (which are all developed countries).

A. Background and goals of the study

In the United Nations publication, International Accounting and Reporting Issues: 1991 Review, it was reported that a comparison in Brazil of environmental reporting by transnational corporations revealed that "...none of the German and Swiss (parent) firms disclosed anything relevant in their reports, in spite of the fact that this information is disclosed in their home countries".

It appeared from the above scenario that transnational corporations with operations in some countries where regulations concerning disclosure of environmental information is somewhat weak may not always disclose information locally even though the parent company may disclose this information in a consolidated format in the home country. This would appear to be because the reporting requirements in the home country are more extensive.

It is the primary objective of this report to examine this observation and presumption in some detail. Environmental disclosures made by transnational corporations operating in selected developing countries, namely India, Malaysia and the Philippines, were analysed. In order to allow for the necessary level of comparability between the different cases, the examples were taken from two industries (chemical and food industries). The corporations were asked to submit for analysis all environmental information which is publicly available in the home and host countries. "Publicly available information" was considered to be any information, whether it is contained within the annual report, an environmental report or is available upon special request.

B. Conceptual background

Corporate environmental reporting is a relatively recent phenomena in the context of environmental management and corporate communications. It is rapidly becoming a primary channel for companies to communicate their thinking, objectives, practices and achievements in the field of environmental management. However, such reporting is still at an early stage of evolution and under the voluntary leadership of a variety of enterprises around the world. In recent years approximately 150 pioneering firms have published environmental reports[1]. These reports vary considerably in terms of scope and quality. Alongside these individual company efforts, some industry associations have begun to issue guidelines on environmental reporting. For example CEFIC, the European Chemical Industry Council, has been encouraging their member companies to publish such reports as soon as possible (more than 50% of CEFIC member companies published such reports in 1993), and it has prepared guidelines to achieve desirable consistency among firms[2].

According to a recent survey on company environmental reporting five levels of corporate environmental disclosure can be identified[3].

(a) Stage 1: "Green Glossies", newsletters, videos, etc.; short statement in annual report;
(b) Stage 2: Special initial environmental report, often linked to a first formal policy statement;
(c) Stage 3: Annual reporting linked to the environmental management system, but more text than figures;
(d) Stage 4: Provision of full TRI-style performance data on an annual basis[4]; input-output data for service companies[5]; corporate and site reports[6]. Information is available on diskette or through online computer systems. The environmental report is referred to in the annual report;
(e) Stage 5: Sustainable development reporting. Aim: no net loss of carrying capacity; linking of environmental, economic and social aspects of corporate performance, supported by indicators of sustainability; integration of full-cost accounting in financial reporting.

C. Approach and methodology

1. Selection of the countries

The project focused on three host countries: India, Malaysia and the Philippines. India was selected because of its economic and political importance in South Asia. Furthermore, it was assumed that the Bhopal disaster had increased environmental awareness and sensitivity in India, especially in the chemical industry. This expectation was confirmed by an international handbook on environmental business strategy: "India's commitment to planned and controlled development and welfare state norms has helped generate concern for the environment. The mining, food processing, heavy machinery and metal fabrication industries in India are leaders in providing information on their environmental performance. Government regulations require companies to disclose in Directors' Reports information on energy conservation and the company's use of pollution control technology."[7] Malaysia and the Philippines were chosen because of their recent economic progress and their role as newly industrializing economies.

Since the mid 1970s all three countries have experienced increased inflows of foreign direct investment. The embassies in Switzerland of all three countries were contacted and asked for information on their regulation of environmental disclosure in their countries[8]. In the case of India local authorities in New Delhi were also contacted. No responses to the inquiries were received.

2. Selection of the industries

The sample of enterprises for this project was chosen from two major global industries: chemicals and food industries. Both are of notable economic and ecological importance. The chemical industry was selected because it has a significant impact upon the environment and therefore enterprises are more likely to disclose environmental information. The food industry was chosen for two reasons: firstly, it was assumed that the production processes are very similar, irrespective of the national cultural or legal framework; secondly, the food industry was regarded as closely related to human health issues.

Therefore, it was felt that there were strong incentives from the marketplace to disclose environmental information.

3. Selection of the corporations

Twenty-six transnational corporations were contacted: 18 companies in the chemical industry and 8 companies were in the food industry[9]. It is hardly possible to make generalizations beyond the sample although the findings are conclusive. Moreover, the sample was likely to be biased because the firms in the sample were regarded as the leading firms with respect to environmental managerial performance.

4. The questionnaire[10]

The questionnaire was designed to capture the type and extent of environmental disclosures made in both the home countries and the host countries. It was sent to the subsidiaries in the host countries. The survey instrument was influenced by the results of United Nations publication, Environmental Disclosures: International Survey of Corporate Reporting Practices (E/C.10/AC.3/1994/4) and the proposals made in the CEFIC guidelines on environmental reporting for the European chemical industry.

The questionnaire consisted of four sections. In the first section the subsidiary was asked in a very general way about publications issued at the local level. Six examples were given: an annual report, an environmental section as part of the annual report, a separate environmental report, press releases on environmental issues, an environmental policy statement, and leaflets on environmental issues. Furthermore, the respondent was encouraged to mention everything else relevant to this topic.

The second section focused on the relationship between the disclosures at the corporate and subsidiary levels. It was designed to determine how information of the local subsidiary in the host country is disclosed at the corporate level in the home country. It was assumed that information concerning environmental matters is often disclosed in a consolidated format in the home country where reporting is more extensive.

The third section provided a structured overview of all kinds of environmental disclosures. Six major categories of different types of information were mentioned and illustrated by examples. The respondent was asked to mark every type of information which was publicly available in the host country, and to submit a concrete example. The types of information listed in the questionnaire were: environmental policy; environmental management; safety; economic and financial information; life-cycle of products and services; and, environmental effects of production activities (data).

In the fourth section the respondent was asked to identify the factors which could account for the current environmental reporting practice. Five factors were listed: legal framework; corporate environmental responsibility of the subsidiary; environmental policy of the parent company; pressure by environmental groups or others such as stakeholders; and pressure from the marketplace. The questionnaire concluded with a strong encouragement to contact the researcher to express further suggestions or questions.

5. Methodology

Initially, the headquarters of 28 transnational corporations in seven countries were contacted. They were requested to send the fax numbers of the subsidiaries in the three host countries and a copy of their most recent annual and environmental report. Twenty-four companies provided some telefax numbers and the publications requested or promised to provide the required information from their subsidiaries in the three countries. Four companies did not respond to the requests.

Secondly, a letter explaining the research project and the questionnaire were sent to 54 subsidiaries in the three countries. Furthermore, a number of responses were reported directly by the parent company.

Initially, the response rate from the developing countries was poor which necessitated follow-up action with both the subsidiaries and parent companies. By the end of project research period responses from 36 subsidiaries within the host countries were received. Nineteen were sent directly from the subsidiary in the host country and 17 were sent by their respective parent company. Of those responding, 11 filled out the questionnaire whereas the remaining 25 preferred to send a letter or informative material only. These 36 responses constitute the sample analysed in this report.

Table IV.1 Subsidiaries contacted

	India	Malaysia	Philippines	Total
Chemical Industry	15 (16) [11]	15 (16)	14 (15)	44 (47)
Food Industry	3	4	3	10
Total	18	19	17	54 (57)

Table IV.2 Details of the country and industry of the respondents

	India	Malaysia	Philippines	Total
Chemical Industry	11	8	10	29
Food Industry	2	2	3	7
Total	13	10	13	36

II. SURVEY RESULTS

The general finding from this project was that environmental disclosures in the host countries is very different from those in the home countries. The overall quantity and quality of the responses was disappointing. In general, environmental disclosure in the home country has reached stages 3 and 4 of the above typology whereas any environmental disclosure in the three countries sampled rarely covers aspects beyond stage 2.

Concerning the sample, two general conclusions can be drawn. The first refers to differences between the two industries. The overall reaction from the food industry was more reserved than the responses from the chemicals industry. In summary, environmental disclosures in the food industry within the host countries could not be evaluated with the exception of replies from Nestlé (India, Malaysia and Philippines) and Pepsi-Cola International (Philippines)[12]. On the other hand, chemical companies could be expected to present some sort of general commitment to environmental

issues (e.g. environmental policy statement). Therefore, a corporation from the chemical industry is more likely to have already taken first steps in environmental disclosure and published at least an environmental policy statement.

The second general finding concerns geographical differences between the three host countries. Subsidiaries in India tended to disclose more information than those in the two other host countries. The following Indian regulations require disclosure of information regarding conservation of energy within the annual report of the local subsidiaries: the Indian Companies Act, 1956 when read with Rule 2 of the Companies Rules, 1988 (Disclosure of Particulars in the Report of Board of Directors) and Section 217 (2-A) of the Companies Rules, 1975 (Particulars of Employees), as amended. Details of the total power and fuel consumption as well as statistics for the various segments (electricity, coal, furnace oil, other generation) are required[13]. Moreover, some Indian companies provide details on energy consumption per unit of production. Assuming that the product mix is constant over a certain period, these figures illustrate the change (improvement or deterioration) of production efficiency as far as energy is concerned.

This legal obligation may have given rise to further environmental disclosures within the framework of conventional financial reporting (annual reports): in the case of Nestlé India and Hindustan Lever Limited a supplementary section on the concern for the environment was included. These sections provide information on various projects and recent environmental improvements.

As a consequence of the limited sample size obtained, the following findings refer to single cases:

(a) Detailed reports providing full TRI-style performance data have become very common for the production sites in the home countries. Nevertheless only one company submitted material which can be regarded as an environmental site report in the narrow sense of the term[14]. The site report of Tioxide (Malaysia) Sdn Bhd, a subsidiary of ICI, consists of seven pages and informs about the following topics:

(i)   environmental policy;

(ii)   the environmental risks of the production process;

(iii)   total waste (data for different categories);

(iv)   by-products;

(v)   environmental management systems;

(vi)   emissions inventory for emissions to air, discharges to water and wastes to land (20 substances);

(vii)   energy use and carbon dioxide emissions;

(viii)   complaints and community relations.

(b)   Several companies submitted explicit environmental policy statements or material which can be interpreted as an environmental policy statement published by their subsidiaries in the host countries[15]. In some cases these statements were translations (English as well local languages) of their corporate policy statements as they had been adopted and reported in the home countries. Some policy statements seemed to be adapted to local needs in the three host countries. In addition to appearing in separate publications, these environmental policy statements were sometimes also integrated with the local annual reports.

(c)   In the case of Ciba-Geigy it was obvious that the type of environmental disclosure at a particular site correlates with the activities carried out at the location. For example, in Malaysia there are no manufacturing facilities. Consequently, local communication tends to be geared more towards involvement in educational and community activities, employee information (which is also available to the public), as well as regular communication with government offices. Similar characteristics apply to the site in the Philippines where some pharmaceuticals production takes place. On the contrary, the facilities in India are much larger and have more production and the information available to the public is structured accordingly. It consists of (among other items):

(i)   An "Our Vision" brochure which is for both internal and external use;

(ii)   "Health and Safety Policy" and "Energy Policy" for India, again for both internal and external information;

(iii)   A brochure about the Santa Monica works which provides information about the business operations as well as activities related to environmental protection, safety

and health;

(iv)   The magazine of the Central Pollution Control Board in New Delhi, which features a photograph of the neutralization system at the Santa Monica site;

(v)   Other information if requested. It is Ciba-Geigy policy worldwide to provide such information unless it is of a confidential nature.

In a number of cases environmental information is available on special request. For example, detailed economic and financial information was disclosed in the questionnaires returned by six firms[16]. According to Bayer Philippines the figures given were not publicly available but may be requested from the Philippine Fertilizer and Pesticide Authority.

As far as information on safety and accidents at work is concerned, Hoechst reveals detailed figures for all three host countries in a leaflet published in the home country (Germany). The leaflet is in the German language and has not yet been translated. It covers all accidents at work worldwide listed by country level.

III. CONCLUSIONS

A number of companies incorporate a pledge to worldwide environmental reporting in their corporate report. Typical phrases include:

(a)   "This report concerns all Rhone-Poulenc businesses worldwide. Every plant must also publish its own year-end results for the benefit of neighbouring communities, as well as its objectives for the following year"[17];

(b)   "Furthermore, all plants (high risk or not) are informing the public about the environmental impact of their activities"[18];

(c)   "It is essential that Monsanto continue to be fully and publicly accountable ... We chose to disclose Monsanto's data on emissions and releases directly to our communities, ... environmentalists and the media."[19]

However, the findings of this report suggest that in some instances companies may not always disclose information on a consistent basis world wide. It must be kept in mind that the whole concept of environmental disclosure is a recent development. It was not until 1990 that the first environmental reports were published and only in 1992 did "Agenda 21", the core document that emerged from the 1992 United Nations Conference on Environment and Development (UNCED), call on businesses and industry to "report annually on their environmental records, as well as on their use of energy and natural resources". Taking these aspects into account, it seems to be reasonable and obvious that a comprehensive environmental reporting system has only recently been introduced in the home countries and the worldwide implementation has not yet been completed.

The difference between the two industry sectors studied points to public exposure as the key factor accounting for different reporting practices. Of course, the legal framework in each country also influences corporate environmental reporting practices. But in the light of the large number of voluntary disclosures it seems to be rather implausible to regard legal forces as a key factor for more comprehensive environmental disclosure. Instead of more stringent environmental regulations a growing environmental concern of the public in the three host countries may contribute to real progress in future environmental disclosures.

Table IV.3 Subsidiaries contacted (chemical industry)

Chemical Industry	Addresses of subsidiaries were provided?	Environmental report from home country available?	Subsidiaries contacted	Subsidiaries responding	Subsidiaries responding with respect to environmental disclosure	Subsidiaries with some sort of environmental disclosure
AKZO	YES	YES	2	0	0	0
BASF	YES	YES	3	2	2	1
BAYER	YES	YES	3	1	1	0
BP	YES	YES	3	0	0	0
CIBA GEIGY	complete information was provided by the parent company in Switzerland	complete information was provided by the parent company in Switzerland	complete information was provided by the parent company in Switzerland	complete information was provided by the parent company in Switzerland	complete information was provided by the parent company in Switzerland	complete information was provided by the parent company in Switzerland
DOW	YES	YES (including European site reports)	3	3	3	3
DU PONT	YES	YES	3	0	0	0
ESSO	NO	NO	0	0	0	0
HENKEL	NO	YES	0	0	The parent company responded to our inquiry	0
HOECHST	YES	YES	3	3	3	3
ICI	YES	YES	3	3	3	2
MONSANTO	YES	YES	3	3	0	0
NORSK HYDRO	YES	YES	3	1	1	0
PROCTER & GAMBLE	YES	YES	3	0	0	0
RHONE-POULENC	YES	YES	3	1	1	0
SHELL	YES	YES	3	1	1	0
UNILEVER	YES	YES	3	0	0	0
UNION CARBIDE	YES	YES	3	2	2	1

Table IV.4 Subsidiaries contacted (food industry)

Food Industry	Addresses of subsidiaries were provided?	Environmental report from home country available?	Subsidiaries contacted	Subsidiaries responding	Subsidiaries responding with respect to environmental disclosure	Subsidiaries with some sort of environmental disclosure
CADBURY SCHWEPPES	NO	NO	0	0	0	0
COCA COLA	NO	NO	0	0	0	0
GRAND METRO-POLITAN	YES	YES	1	0	0	0
HEINEKEN	YES	YES	2	1	1	0
KRAFT GENERAL FOODS	YES	YES	1	1	1	1
MC DONALDS	NO	NO	0	0	0	0
NESTLE	YES	YES	3	3	3	3
	complete information was provided by the parent company in Switzerland	complete information was provided by the parent company in Switzerland	complete information was provided by the parent company in Switzerland	complete information was provided by the parent company in Switzerland	complete information was provided by the parent company in Switzerland	complete information was provided by the parent company in Switzerland
PEPSI	YES	YES	3	2	2	1

Table IV.5 Survey results (chemical industry)

Chemical Industry	Number of cases	Cases reported by the headquarters	Cases reported by the subsidiary	Returned questionnaires
AKZO	0	0	0	0
BASF	2	0	2	0
BAYER	3	2	1	1
BP	0	0	0	0
CIBA GEIGY	3	3	0	0
DOW	3	3	0	0
DU PONT	0	0	0	0
ESSO	0	0	0	0
HENKEL	2	2	0	0
HOECHST	3	0	3	3
ICI	3	1	2	1
MONSANTO	3	3	0	0
NORSK HYDRO	1	0	1	0
PROCTER & GAMBLE	1	0	1	0
RHONE-POULENC	1	0	1	0
SHELL	1	0	1	1
UNILEVER	1	0	1	1
UNION CARBIDE	2	0	2	1

Table IV.6 Survey reults (food industry)

Food Industry	Number of cases	Cases reported by the headquarters	Cases reported by the subsidiary	Returned questionnaires
CADBURY SCHWEPPES	0	0	0	0
COCA COLA	0	0	0	0
GRAND METROPOLITAN	0	0	0	0
HEINEKEN	1	0	1	1
KRAFT GENERAL FOODS	1	0	1	0
MC DONALDS	0	0	0	0
NESTLE	3	3	0	2
PEPSI	2	0	2	0

Annex I

QUESTIONNAIRE ON ENVIRONMENTAL DISCLOSURES

1. Do you publish at a subsidiary level:

   - an annual report?
   - a section on environmental issues as part of your annual report?
   - a separate environmental report?
   - press releases on environmental issues?
   - leaflets on environmental issues?
   - an environmental policy statement?
   - anything else (please explain)?

2. Is your environmental information (data, expenditures etc.) integrated into a corporate environmental report of the corporate annual report of your parent company?

3. What types of environmental information are publicly available? Please send us an example.

a) Environmental policy

   - View on environmental demands
   - Principles
   - Targets (qualitative or quantitative objectives)

b) Environmental management

   - Environmental management systems
   - Major programmes
   - Environmental audits
   - Human resources, training and educations
   - Environmental officer and health safety staff

c) Safety

   - Accidents
   - Injuries
   - Risks/environmental risk assessment
   - Emergency preparedness
   - Lost working days

d) Economic and financial information

   - Total expenditures
   - Capital expenditures
   - Operating costs
   - Remedial expenditures
   - Research and development expenditures
   - Liabilities

e) Environmental life-cycle of products and services

   - Energy
   - Resources
   - Emissions
   - Waste

f) Environmental effect of production activities (data)

   - Energy generation and consumption
   - Resource consumption data
   - Emissions data
   - Waste and by-products

4. What factors do account for your environmental reporting practice?

   - The legal framework in your host country?
   - The corporate environmental responsibility of the subsidiary?
   - The environmental policy of your parent company?
   - Pressure by environmental groups of other stakeholders?
   - Pressure from the marketplace?
   - Other factors (please explain)?

In addition to this questionnaire we would like to receive samples of your environmental reporting.

Annex II

PROPOSED COMMON STRUCTURE FOR SITES ENVIRONMENT REPORTS PREPARED BY THE EUROPEAN CHEMICAL INDUSRY COUNCIL (CEFIC)

When environment reports are established or individual sites of a company, it is recommended that they be drafted along the following headings:

1. Forward

- Site manager address
- Company environment policy
- Company environmental objectives (medium-long term)

2. Site description

- Main units, main products
- Site put into perspective - usage of products
- economic contributions and employment
- relations with authorities and local community
- Environmental situation: local conditions of air, water etc. in the neighbourhood, sensitive areas, etc.
- Legal requirements: permits, emission limit values, etc.
- Controlling authorities (national, local etc.)

3. Environmental management

- Structure (human resources and organisational)
- Programmes and objectives
- Environmental protection techniques (water treatment, waste incinerator, waste minimization, etc.)
- Integrated approach (recycling and new technologies)
- Monitoring techniques/systems (data measured/calculated/estimated)
- Emergency plan

4.Data (with comparison with data on previous years)

- Emission data
- Selected details (noise, odour etc.)
- Health and safety data

- Complaints (optional)
- Spending on environmental protection

5.Communications

- Community relations
- Open days

6.General comments

7.Contact people

Note: Information should be adopted to site specifications and local expectations.

Notes

1.   According to a recent survey by a German research team 127 companies have published an environmental report (Clausen/et.al. (1993), p. 10). A recent report by UNEP evaluates the environmental reports of 100 environmental reporting pioneers (UNEP Industry and Environment Program (1994), p. 5).

2.   CEFIC (1993).

3.   UNEP Industry and Environment Program (1994) p. 28.

4.   TRI stands for Toxic Release Inventory. In the United States companies report annually certain TRI pollution data for more than 300 chemicals.

5.   Environmental reports at this stage cover the range of substances going through the entire production process (from the input to the output).

6.   It is crucial to distinguish between site and corporate-wide reporting. A site report deals with the environmental effects of a particular production site, whereas a corporate environmental report aggregates the information of a number of sites at the corporate level.

7.   IISD (1992), p. B91.

8.   cf. Annex I.

9.   Tables 3 and 4 provide information about the selected transnational corporations.

10.   cf. Annex II.

11.   This number includes the cases directly reported by the parent company.

12.   There were three other subsidaries responding to the inquiry but none of them submitted any information on environmental disclosures. In the case of Kraft General Foods Philippines the response only referred to the very general information previously submitted by the parent company. For Pepsi-Cola, the results are significant: Pepsi Cola International (Philippines) submitted a wide range of information: the environmental effect of production activities (wastewater characteristics), capital expenditures on wastewater treatent, environmental policies and environmental management systems including goals and action plans. In contrast to Pepsi Cola International (Philippines), the Indian branch, Pepsi Foods Limited in New Dehli, reported that the "...activities in which [they are] presently engaged [are] of a non-polluting nature." For this reason they "...do not publish any report on environmental issues/information." The third case comes from a subsidary of Heineken that stated that the survey did not apply to their company.

13.   This is the legal norm for all annual reports filed with the Indian government. Consequently a comparable section on the conservation of energy was available from BASF India Limited, Bayer (India) Limited, Hoechst India Limited, Nestlé India Limited and Hindustan Lever Limited.

14.   A comprehensive environmental site report should provide a full picture of its environmental impact. A structure coming up to this expectation was proposed by the European Chemical Industry Council (CEFIC) in June 1993 (cf. Annex II).

15.   Seven responses were received which included at least aspects of an environmental policy statement: Dow Pacific, answering the inquiry to their offices in the three host countries, sent an English version of their corporate environmental policy; the Indian and Malaysian subsidaries of ICI PLC submitted an environmental policy statement published by the local subsidaries. The same occurred for Ciba-Geigy India, Hoechst Philippines Inc., Hindustan Lever Ltd. and Nestlé (Malaysia).

16.   Hoechst India, Hoechst Malaysia, Hoechst Philippines, ICI India, Nestlé India, and Union Carbide Philippines.

17.   Rhône-Poulenc (1993), p. 3.

18.   Rhône-Poulenc (1993), p. 24.

19.   Monsanto (1993), p. 3.

References

CEFIC, 1993, <u>CEFIC Guidelines on Environmental Reporting for the European Chemical Industry</u>. Brussels.

Clausen, J. and K. Fichter, 1993, <u>Vorstudie zum Projekt Umweltberichterstattung</u>. Berlin.International Institute for Sustainable Development, 1992, <u>Business Strategy for Sustainable Development: Leadership and Accountability for the '90s</u>. Winnipeg, Canada: IISD.

Monsanto Chemical Co., 1993, <u>Environmental Annual Review 1993: Building a Sustainable Future</u>.

Rhône-Poulenc, 1993, <u>Environment Report 1993</u>.

United Nations, 1994, <u>Environmental Disclosures: International Survey of Corporate Reporting Practices</u>, E/C.10/AC.3/1994/4.

United Nations Environment Programme, Industry and Environment Programme Activities Centre, 1994. <u>Company Environmental Reporting</u>, Technical Report Series No 24. Paris, United Nations.

Chapter V

A PRACTICAL APPROACH TO INTEGRATED ENVIRONMENTAL ACCOUNTING: AGIP PETROLI CASE STUDY

Summary

The Fondazione Eni Enrico Mattei (FEEM)[1] has developed a methodology for preparing corporate environmental balance sheets that aims to satisfy both private and public needs for comprehensive information on company flows of inputs, outputs and monetary values related to environmental issues. The balance sheets have a flexible structure that can be shaped and aggregated at different levels: a single plant, a production line, a site, a division, a company, a group of companies or a defined geographic region. Because of this feature, the balance sheets are useful for both private and public decision-making. From a company perspective, these environmental balance sheets can be useful: to identify causes of the inefficient management of natural resources; to simulate future situations; and to plan and manage measures to improve environmental performance. Avoiding double counting, the methodology allows different consolidations: the analyst can produce a balance sheet for a group of companies that is similar to a financial consolidated balance sheet. From a government perspective, this structure is useful to understand at local or national levels what is the pressure of a specific industrial sector on the environment and the financial resources needed to reduce that pressure. In fact, it is possible to calculate the balance sheet for a geographic area by making consolidations of different units, sources of pollution and resources used by consumers.

The balance sheet is not just theoretical: it has been implemented in some large companies

operating in Italy. This paper focuses on one of these companies, Agip Petroli, to demonstrate a clear example of the usefulness of the methodology as a management tool and as a provider of information to external users.

INTRODUCTION

The environmental balance sheet is an information instrument theoretically similar to the financial balance sheet. While the latter contains economic values, the former collects and rationalises relevant data for environmental management.

Compared with traditional accounting practices, environmental accounting does not have a formal equivalent of the balance sheet on which executives of companies should base their decisions. Investments for environmental management are still planned on the basis of the emotional perception of problems without a consistent framework to support decisions.

This paper suggests an answer to company information requests on environmental issues. The proposed balance sheet structure (now operationalised in a computer software package) has been developed by a FEEM research team that has been working for several years in close cooperation with companies, studying their decision-making processes and their lack of information related to environmental issues. This collaboration has made possible the formalisation of a balance sheet structure that is an efficient provider of all of the relevant quantitative information that is necessary for effective environmental management: physical flows and monetary values of input and outputs and data on expenditures for environmental protection.

This methodology is also the result of a joint project between FEEM and ISTAT (the national Italian statistical office) which was started in 1991 that is concerned with the development of a system of national environmental accounts to provide policy makers with information on monetary values and flows of inputs and outputs at an aggregate level. The objective of the FEEM-ISTAT project is to set conditions and guidelines for the determination of a "green" Gross Domestic Product (GDP).

I. BALANCE SHEET STRUCTURE

What is an environmental balance sheet? Researchers as well as managers do not have the same definition of its contents and use this term for a wide range of instruments that collect and communicate data.

In Germany and Switzerland, the so called "Okobilanz" has gained a certain success as an environmental information tool. The Okobilanz is a sort of mass balance sheet which collects physical data on the consumption of raw materials and emissions and, in some cases, these figures are translated into eco-points (a measure of the environmental impact of some emissions). The Okobilanz approach, even if very interesting especially when eco-points are calculated, does not integrate decisions affecting the environment with traditional decision making process.

Starting from this weak point of traditional data recording and more innovative approaches, it was necessary to find a way to connect the environmental dimension with the economic sector. The FEEM methodology suggests such an integration by collecting physical data and associating it with economic values. This information is generally addressed separately and not comprehensively. In particular, the methodology involves a collection and analysis of:

(a) physical data on inputs and their related economic value;

(b) physical data on products and their related economic value; and,

(c) physical data on emissions and their related environmental expenditures.

The methodology tries to link the three accounting sections in order to be able to assess the economic impact of pollution control and for managing the efficiency of raw materials consumption.

A. Physical accounts for inputs and their related economic value

According to the suggestions of the United Nations Statistical Office (UNSO, <u>Guidelines for Physical Accounts</u>) and "S.E.R.I.E.E." (Système Européen de Rassemblement de l'Information Economique sur l'Environnement), the accounts for physical inputs are divided into two sections:

(a) quantities and values of products purchased for the site, company, or corporation in the market place or coming from some production process (produced goods); and

(b) quantities and values of non-produced commodities, the resources directly taken from the natural heritage.

To clarify the distinction, water can be considered as an example of an input. If the water is taken directly from a river or a lake (and without paying for it), it has to be included among the non-produced goods. On the other hand, if the water is taken from the municipal utilities network it must be considered to be a commodity. Again, considering oil consumption, crude oil is a non-produced good for the exploration company and a commodity for the downstream oil company.

In order to properly use the information on inputs, it would be useful to gather data relating to: country of origin (for the purpose of evaluating the impact of the company on the national natural heritage), supplier (to avoid double accounting in the company consolidated balance sheet, because if the product comes from a company belonging to the same group it does not need to be accounted for again) and price (for example, this information is useful for determining the cost for the company to shift from resources which pollute to environmental-friendly resources).

This information is very useful: to avoid double counting in a hypothetical consolidated environmental balance sheet; to permit an analysis of input flows in a consolidated environmental balance sheet; and to focus on the efficiency of the company in transforming a non-produced good into a commodity.

B. Physical accounts for products and their related economic values

Similar to the accounts for inputs and their related economic values, these account register the quantity of products in the maximum detail that it is possible to obtain. In particular, each kind of product has to be registered according to its final destination (for the purpose of making consolidations) and its value. Other information on the quality of products can be registered: for example the content of toxic pollutants that the company wants to monitor during all of the life cycle. The information on total quantity and value of sales is then used to calculate environmental performance indicators that use these figures as denominators.

C. Physical accounts for emissions and their related environmental expenditures

Company outputs also include emissions into the air, water discharges, wastes and noise. This account records quantities emitted in different media and links them with environmental expenditures.

While the accounts of inputs and products contain information on economic values (the price of raw materials and the value of sales) for the different kinds of emissions the integration between the environmental and economic sectors is not as evident.

Researchers for a long time have been trying to find a rational and feasible way to translate emissions into impacts, and impacts into monetary values of environmental damage. Although the theoreticians are making some important progress, there is not an accepted methodology for evaluating the externalities related to industrial pollution. Therefore it is necessary to find practical approaches (both for companies and for governments) to include environmental protection into traditional decision-making by finding a measurement that can explain the value of environmental protection measures taken by a company.

Environmental expenditures are determined for purposes of preparing the FEEM balance sheet in a way that links the environmental and the economic conceptual spheres. The objective of this approach to data recording is firstly to control investments and operational expenditures in relation to their effectiveness. But, as the practice of registering environmental expenditures becomes a normal operating procedure in a company, information on the levels of emissions and the company efforts to reduce those levels represent a valid support in the decision-making process to plan action steps, investment controls and environmental budgeting programmes.

At the national level, the aggregate account of environmental expenditures is necessary information for policy makers and gives an indication of the amount of money spent by the production sector for environmental protection. The connection of this methodology with the national statistical accounts is ensured by the classifications contained within "S.E.R.I.E.E.".

D. The problem of determining the classification of environmental expenditures

According to the European Statistical Office (EUROSTAT, 1992), environmental expenditures are only those deliberately and mainly undertaken to prevent, control, reduce or eliminate the negative effects on the environment, without considering whether or not they are required by legislative requirements. EUROSTAT does not consider environmental expenditures to include those that are made for the purpose of cost savings, technical requirements, labour safety or health of personnel, even if they have a positive effect on the environment. In this methodology, "Health, Safety and Environment" (HSE), that is generally considered as one activity, are not accounted for together, and this scheme of preparing an environmental balance sheet includes only environmental protection measures.

Theoretically, accounting for environmental expenditures would seem to be an easy procedure if the accounting system is properly set up; but a major difficulty arises because in many situations it is not easy to draw the boundaries between the expenditures undertaken to gain economic benefits and those undertaken to reduce pollution. Especially when the decision to invest money is voluntary, expenditures are usually planned both for pollution prevention and for technical and economic purposes (see Figure V.1 for a schematic description of the calculation process).

In the FEEM methodology and according to EUROSTAT procedures (EUROSTAT, 1992), environmental expenditures are divided into eight main categories: air and climate protection, water and non-underground water protection, wastes, soil and underground water protection, noise reduction, natural heritage, research and development expenditures, and expenditures for other environmental protection activities (including training and administrative costs).

Idealistically, the target is to relate expenditures to each source of pollution, linking for example, the quantities of emissions of CO_2 gases to the capital and current period expenditures to reduce it. Therefore the categories of air, water and wastes are shared in as many columns as the number of pollutants are released into the media. This is not a feasible approach because there are very few investments undertaken to reduce just one pollutant. More practically it will be necessary to create some groups of pollutants to which can be imputed the cost of a single environmental expenditure; in this way, for example, the expenditures for a filter in a smoke stack will contribute to the reduction of many pollutants.

Each organisation, or level of organisation, can also decide to account for environmental expenditures using more than one classification: according to environmental legislation (voluntary and compulsory), according to the objectives of the expenditures (preventive or reparation), and according to the economic function (current and capital expenditures).

The FEEM methodology considers specifically the third classification leaving the others to the environmental policy of the organisation. This means that companies which want to follow this scheme must adopt at least the classification between current and capital expenditures; subsidiary classifications will depend on the objectives of the company's own accounting system.

This classification has been chosen as the most significant following the guidelines in environmental accounting set by many international accounting bodies now concerned with environmental issues (e.g. the United Nations Conference on Trade and Development and its Intergovernmental Working Group of Experts on International Standards of Accounting and Reporting).

II. THE AGIP PETROLI CASE STUDY

A. The company in brief

Agip Petroli is the main downstream oil company operating in Italy in the fields of: procurement of crude oils, refining and distribution of petroleum products, supply of services for energy savings and for the rationalisation of consumption. It accounts for over 50% of the Italian market and in 1993 Agip Petroli and its subsidiaries had total sales of around 45,000 billion Lire and investments of 1,186 billion Lire. It belongs to the ENI Group, a publically-owned corporation dealing in many strategic sectors such as energy, chemicals, engineering, industrial equipment and machinery.

Figure V.1. *How to calculate environmental expenditures*

Figure V.1. How to calculate environmental expenditures

```
COMPANY ACTIVITY THAT HAS
POSITIVE EFFECTS ON THE ENVIRONMENT

Is the main objective of the activity the protection of the environment?

YES - all the costs have to be identified
NO - The expenditure is not environmental but can be accounted in a separate account

Is the activity carried out in the company?

NO - It is purchased from external companies: the expenditure is environmental
YES - Is the activity carried out by business unit only involved in environmental management?

NO: environmental protection activities are integrated with other production activities
YES: current and investment expenditures for the unit are environmental

Is it possible to divide between the environmental part of the activity and the other company activities?

YES: environmental measures are end of pipe. The expenditure related to environmental part of the activity is environmental
NO: environmental protection activities are integrated to production cycle

Do environmental protection measures require a modification of existing processes or operations?

YES: the modification is environmental
NO: environmental protection measures are integrated with a new plant. The environmental expenditure has to be estimated using:
1) additional cost methodology comparing the cost of the plant with a plant with lower environmental performance
2) other estimation methods of the environmental part of the measure
```

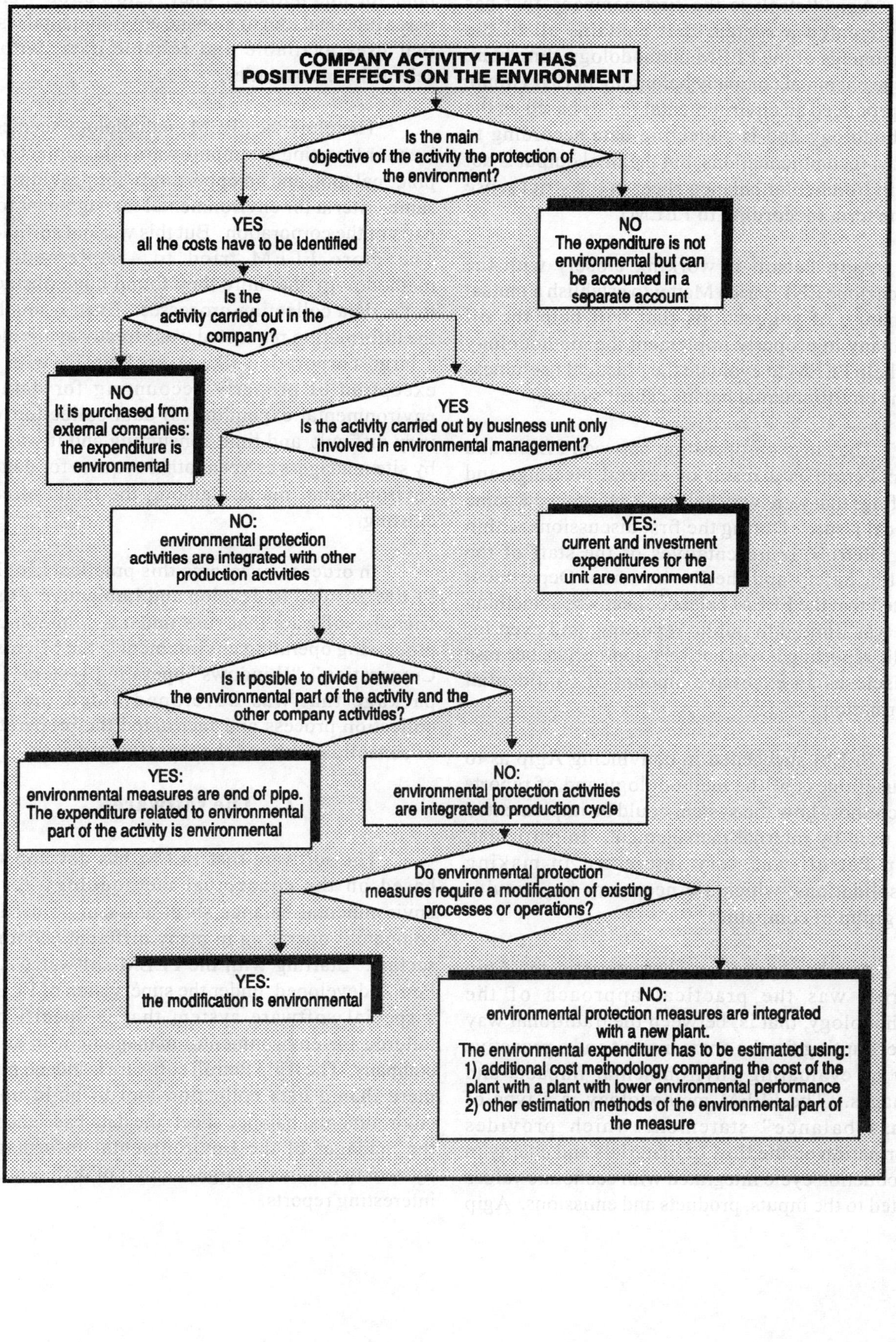

B. The feem balance sheet at AGIP Petroli

Agip Petroli is the first company that has implemented a system that contains all of the components of the FEEM methodology. Enichem, a large chemical company belonging to ENI Group, has produced an environmental report based on the methodology and is collecting data according to FEEM recommendations, but the final output is not an environmental balance sheet in the strict sense of the term (according to FEEM).

Agip Petroli is working closely with the Fondazione ENI Enrico Mattei to establish a mutual exchange of suggestions that will help the oil company to properly implement the methodology and help FEEM to evaluate the practical usefulness of the environmental balance sheet system.

The close co-operation between FEEM and Agip Petroli is a result of several meetings and briefings that were very helpful in pointing out some critical issues. During the first discussions within Agip Petroli, representatives of the staff of the Health, Safety and the Environment Department argued that the FEEM balance sheet was something that was limited to public reporting, and even for public reporting it was not very appropriate because it discloses an excessive amount of confidential information.

FEEM succeeded in convincing Agip as to the usefulness of the methodology and of reports which show how the system could be implemented and used by internal management. In particular, Agip Petroli was very interested in making consolidations, which are necessary for a multi-site group of companies.

Another appealing characteristic for Agip Petroli was the practical approach of the methodology, that is, between the traditional way of accounting for the environment and a proactive way of evaluating environmental impacts and damages. The FEEM methodology is a type of "mass balance" statement which provides information on the flow of materials and energy in a production cycle integrated with economic values related to the inputs, products and emissions. Agip Petroli was not at the stage of making a monetary evaluation of environmental damages and also their traditional accounting system seemed extremely poor for this purpose: what Agip Petroli wanted was a practical tool to manage environmental issues that uses available and reliable data, without scientific uncertainties.

Therefore the FEEM methodology seemed to be an interesting attempt towards this objective, a practical tool, and an opportunity for providing the same criteria for environmental management to all parts of the corporation. But this was not sufficient. The more FEEM tried to implement the methodology the more they found huge obstacles to the data collection processes. Even if much of the information requested was already available in a large corporation like Agip Petroli (with the exception of properly accounting for data on environmental expenditures) data collection was very difficult and time consuming and it was felt by site managers as yet another request for data on environmental matters among the many already existing.

In order to overcome this practical problem FEEM decided to develop and implement a user-friendly software that runs under a common data processing operating environment -- the Microsoft Corporation's "Windows" operating system. The package that was developed aided the data collection process and facilitated the preparation of consolidations and the reporting process.

C. The software

The software that FEEM has developed is based on the conceptual methodology for an environmental balance sheet and can be used by companies operating in many different economic sectors. Starting with the FEEM software, Agip Petroli developed, under the supervision of FEEM, a special software system that is specifically tailored for environmental management in an oil company. The Agip Petroli software is indeed much more than a data collection tool to build up an environmental balance sheet: the data base enables the manager of the Environmental Information System to produce a large number of different and interesting reports.

Figure V.2: ***The complete environmental balance sheet***

THE COMPLETE ENVIRONMENTAL BALANCE-SHEET

	RESOURCES								GOODS AND SERVICES			POLLUTANTS AND EXPENDITURES																											
Data collection units	Produced goods				Non-produced commodities							Air						Water						Soil, underground and underground waters						Wastes						Noise		Current expenditures	Investments expenditures
	1				2							Group of pollutants 1				*Tot ce*	*Tot ie*	Group of pollutants 1				*Tot ce*	*Tot ie*	Group of pollutants 1				*Tot ce*	*Tot ie*	Group of pollutants 1				*Tot ce*	*Tot ie*	ce	ie	ce	ie
	1.1		1.2		2.1		2.2					1.1	1.2	sc	si			1.1	1.2	sc	si			1.1	1.2	sc	si			1.1	1.2	sc	si						
	q	v	q	v	q	v	q	v	q	v	av	q	q	v	v	v	v	q	q	v	v	v	v	q	q	v	v	v	v	q	q	v	v	v	v	v	v	v	v
Unit 1																																							
Unit 2																																							
Unit 3																																							
Company (total)																																							

Fondazione ENI Enrico Mattei

Symbols:

1,2 code of the family of product or pollutant
1.1, 1.2, 2.1, 2.2 = code of the single product or pollutant

ce = current environmental expenses
ie = investment expenses

v = monetary value
q = physical quantity
av = added value

The package creates a data base that can be used and aggregated according to different criteria. Figure V.3 shows the structure of the software that consists of data entry, an accounts section and different kinds of data aggregation for reporting purposes.

1. Data entry

One of the main purposes of the software is to simplify the data entry process and to keep it homogeneous during a time period and among different sites. In fact, a multi-site company like Agip Petroli has had in the past (and partly still has) tremendous problems using the same classifications for production input, output and pollutants: the best way to standardise the information seemed to be a software package wherein the user nearly always has to choose among pre-determined lists of alternatives.

Also, the huge amount of data requested by the Environmental Information System requires the use of computers to correct mistakes, to integrate

incomplete information and to make calculations. Some examples of the kind of data requested during the data entry process can help to understand the issues involved. The data entry section is divided into:

(a) consumption of raw materials;
(b) goods and services produced by the unit;
(c) wastes;
(d) noise;
(e) air emissions (excluding non-point sources);
(f) non-point emissions;
(g) water discharges; and,
(h) environmental expenditures.

For each of the above topics, many kinds of information are requested. For example, the information on the consumption of raw materials needs to include:

(a) the code of the raw material (the list of codes is taken from the Agip Petroli accounting system);
(b) the name of the raw material (the list of names is taken from the Agip Petroli accounting

Figure V.3: *The structure of the Agip Petroli Environmental Information System software*

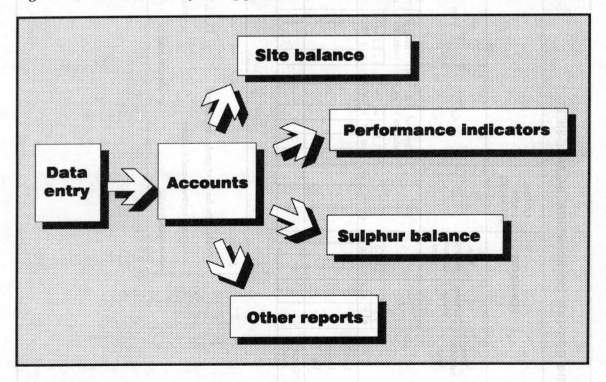

system); for a particular kind of crude oil, for example, the software shows default figures for the content of sulphur, nickel, vanadium, benzene, aromatics, etc.;

(c) the price (this information is taken from the Agip Petroli accounting system);

(d) the country of origin (the objective of this information is to distinguish between different kinds of local environmental conditions); and,

(e) the supplier (for consolidation purposes, if the supplier belongs to Agip Petroli the user has to fill in a field of information with the name of the company, taken from a specified list).

More complex is data entry for air emissions (excluding non-production point sources). This topic was addressed following very specific requests by Agip Petroli. For each kind of air pollutant the user has to compile a schedule that includes:

(a) for equipment: the name and code number of the equipment, its power requirements, the kind and quantity of fuels used, the height and diameter of the plant chimney, and the number of operational hours of the equipment;

(b) for the plant: the name and code number, its capacity, and its operational hours of the equipment; and,

(c) for pollutants: concentrations, quantities, kinds of abatements, objectives of reduction (both concentration and quantity), legal requirements, and deadlines for reductions to occur.

The data entry process is assisted by formulas and utility procedures with a frequent use of codes that optimise the input time required.

2. Accounts

During data entry, records are registered by the computer in chronological order: therefore, for example, data on SO2 emissions are not added to each other even for the same site.

The second step of the Environmental Information System adds records and deletes some kinds of information that will not be relevant on an aggregated basis. Accounts for each of the previous topics (consumption of raw materials, goods and

services produced by the unit, etc.) are prepared on a site level. In this section a quantification of sulphur content is produced, taking into account:

(a) sulphur coming into the process, typically in crude oil; and

(b) sulphur coming out from the process: in air emissions (SO2) both from the local plant and non-plant sources, in products (for each product the data entry controls require that the content of sulphur and other important pollutants be specified) and in pure sulphur sold as a product.

3. Site balance sheet

The site balance sheet is one of the outputs of this sophisticated system implemented by Agip Petroli. The format of the balance sheet gives a detailed picture of the situation at the refinery with the main issues expressed both in economic and physical values. Comparing the format of this balance sheet with those of other well-known oil companies, it appears that this format has the capabilities of being able to present in one report information that is usually contained in several reports.

Actually, some external environmental reports have extensive data and tables, but the reader has to find the desired information by reading documents that contain thirty pages or more. This balance sheet includes all of the sections that are relevant from an environmental point of view. Sections on resources, products and emissions are linked together by economic data that allows for comparisons and normalisations (see Figure V.4).

From an internal point of view, this balance sheet can be addressed to top management periodically, preferably with a special section on performance indicators that summarises and compares the current situation with previous and future objectives. The usefulness of the balance sheet is not limited to its efficient format for presenting data. The information can be divided and consolidated in many ways to provide information on the reasons of the trend of emissions for use in controlling investment decisions.

The balance sheet is also useful for benchmarking policies. This implementation strategy focuses on a single site, but in the coming future it will be carried out on other refineries and facilities: this process will enable top management to control and compare the performance on different sites that are linked to other economic information.

4. Consolidation

Currently it is possible to aggregate data to make consolidations. Other reports are produced at this stage. After the preparation of the balance sheet, each site sends data to the operating level above its own (in a specific corporate structure) with a special format. Of course, in going up to higher levels in the corporate structure there is a necessary loss of detailed information: only the information required for the consolidated balance sheet and other special reports is transmitted. The consolidation is necessary in a group like Agip Petroli because at different levels there is a need for knowing the relationship between the corporation as a whole and the environment.

Looking at Figure V.5 the consolidation can be done at the level of Refinery 2 (if Refinery 2 has at least one level below it) or at the total Agip refinery operations level, or at the Agip Petroli level. In fact, in order to build up the consolidated balance sheet at the holding company level, each refinery has to build up its own balance sheet, prepare data for the aggregation, and send data to the next higher level. The level above does the same, and so on up to the Agip Petroli Group level.

The usefulness of the consolidation is clear: it avoids double counting, especially relating to refineries, storage facilities, and lubricants plants, and it enables top management (at different levels) to understand the relationship with the environment of all areas below by considering the area as a single unit that has physical and monetary exchanges with the external environment.

The consolidation is not limited to the Agip Petroli organisational structure: in fact the manager of the Environmental Information System (at the holding company level) can make consolidations using any group of entities that is desired. For example it is possible to prepare a consolidated balance sheet for refining facilities or lubricant plants or storage facilities. Another criteria could be the size of the site or a geographical location preparing consolidations, for example, for northern and southern sites.

5. Environmental performance indicator

Starting with the data base created in the Environmental Information System, it is possible to calculate the most useful environmental performance indicators for a single refinery or for the entire Agip Petroli organization. The figures to build up performance indicators are the same as in the balance sheet: raw materials and energy consumption, products (quantity and value) and emissions in different media.

In order to develop comparable and significant performance indicators for Agip Petroli refineries, raw materials consumption and tonnes of emissions are divided by the value added, total sales and earnings and crude oil throughput (as a measure that describes the level of activity of the site). Environmental performance indicators enable managers to set objectives and monitor their achievements and to compare the efficiency of the use of resources in different years and among different sites.

The comparison among sites has proved to be very useful as a benchmarking tool: management at each site can compare their performance related to individual issues of environmental protection with the performance at other sites and therefore exchange positive and creative solutions for improvements.

6. Sulphur balance

One of the most stringent problems for the downstream oil industry is represented by sulphur dioxide (SO_2) emitted from refineries and as contained in products. Legislation is becoming more stringent and imposes on companies the requirement to control the sulphur cycle inside the company and to exchange it with external agents.

Figure V.4: *The complete environmental balance sheet of a refinery*

	1991		1992	
	Quantity (Tons and m3)	Values (Lit * 1000)	Quantity (Tons and m3)	Values (Lit * 1000)
RESOURCES				
Crude oil (tons)	2.850.000	418.036.300	3.219..000	507.700.000
Lead (tons)	62	560.952	66	592.920
Sea Water (m3)	50.465.983	0	57.000.000	0
River, Lake Water (m3)	1.505.126	0	1.700.000	0
PRODUCTS				
LPG	69.413	17.360.413	78.400	19.947.232
Naphta	30.191	6.221.624	34.100	7.148.688
Premium Gasoline	311.640	98.095.740	329.400	105.479.161
Unleaded Gasoline	209.487	70.424.884	259.200	88.644.067
Kerosene	21.754	6.259.032	25.700	7.522.262
Automotiove Diesel	1.142.151	333.868.001	1.288.900	383.280.731
Fluid Fuel Oil	100.427	16.517.900	112.300	18.790.194
Fuel Oil ATZ (High S)	82.225	8.342.415	94.000	9.702.069
Fuel Oil PTZ (Low S)	630.783	90.744.164	718.100	105.092.140
Bitumen	104.781	15.034.082	112.700	16.449.960
Liquid S	9.208	416.911	10.400	479.034
TOTAL PRODUCTS (Tons & Values)	2.712.060	663.285.166	3.063.200	762.535.538

	1991 Quantity		1992 Quantity		ENVIRONMENTAL EXPENDITURE 1992		
					Current	Investm.	Total exp.
POLLUTANS						Million Lit.	
AIR							
CO2	750.000		780.000		0	0	0
SO2	4.500		4.100		600	950	1.550
NOx	1.400		1.420		0	0	0
VOC	1.880		1.500		55	400	455
PST	180		183		20	10	30
Total	757.960		787.203		675	1.360	2.035
WATER							
BOD	7.100		7.500				
COD	56.300		60.000				
SST	27.000		30.000				
OILS	1.300		1.500				
FENOLS	140		150				
N(NH4)	2.600		3.000				
Total	94.440		102.150		1.250	250	1.500
WASTES							
Munic.	200		190		25		25
Special	1100		1000		430		430
Toxic and Hazardous	1,4		1,5		0		0
Total	1301,4		1191,5		455		455
					1.705	250	1.955

Figure V.5: *Information flows for the Environmental Information System*

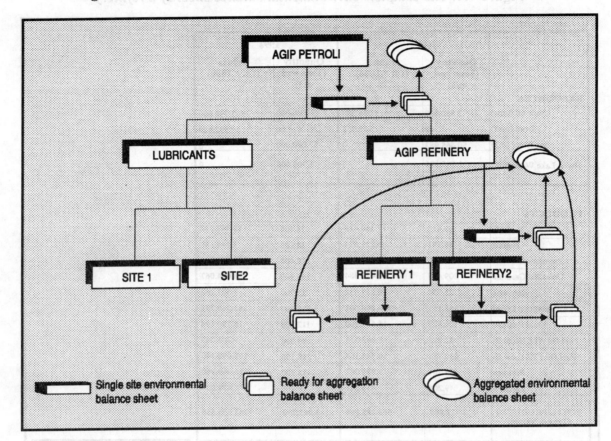

Agip Petroli decided to exploit the software in order to calculate automatically the sulphur balance by making calculations of existing data in the Environmental Information System. The result is a schedule that shows:

(a) the sulphur coming into the production process as contents of the crude oil;

(b) the sulphur emitted as SO2 emissions;

(c) the sulphur sold as a specific product; and,

(d) the sulphur output of the production process as contents of products.

The data on sulphur quantities is useful both for monitoring and for planning investments. Since environmental performance for sulphur control is related to different areas (raw materials and product quality, "end of pipe" and integrated measures) its improvement depends on large investments throughout the entire production process.

7. Other reports

Once data are collected in a consistent and significant way, there are virtually no limits to the kinds of information that the manager can produce in specific reports. At the moment, the system has only recently been implemented and only certain specific reports have been prepared.

A report, called the "shuttle account", puts in relationship environmental performance in two or more years with the expenditures to protect the environment. Figure V.6 is an example of the shuttle account where environmental emissions (air, water and soil) are compared with data on sales, value added and quantity of production in two different years (1993 and 1994 in this example). The purpose of relating this data to environmental expenditures is to assess the effectiveness and efficiency of such expenditures in reducing pollution.

Another report that is very useful at Agip Petroli is related to air emissions whereby it is possible to compare the performance of the same kind of equipment at different sites. Information can be presented at the level of individual smoke stacks (production units) and summarized up to total plant levels.

For external purposes, different from the balance sheet in itself, there are many interesting implementations and improvements opportunities for the system to prepare reports for public authorities using the same data base as the company uses.

Figure V.6. *The Shuttle Account*

Kind of Pollutant	ENVIRONMENTAL EXPENDITURES				ENVIRONMENTAL PERFORMANCE INDICATORS					
	1993		1994		1993	1994	1993	1994	1993	1994
	Current Expenditures	Investment Expenditures	Current Expenditures	Investment Expenditures	Pollutants/ Sales		Pollutants/ Added Value		Pollutants/ Quantity of Production	
Air SO2 NOX PST										
Wastes Special Toxic										
Water BOD COD Oil NH4++										
....										

References

EUROSTAT, 1992, Système européen de rassemblement de l'information économique sur l'environnement-S.E.E.R.I.E. Luxembourg: EUROSTAT.

GEMI (Global Environmental Management Initiative), 1994, Finding Cost-Effective Pollution Prevention Initiatives: Incorporating Environmental Cost into Business Decision Making. Washington: GEMI.

Gray, R.H., 1993, Accounting for the Environment. London.

Peskin, H.M., and Lutz E., 1990, "A Survey of Resource and Environmental Accounting in Industrialised Countries", Environmental Working Paper #37. Washington: The World Bank.

Sammarco, G., 1993, "in Ambiente e Contabilità Nazionale", Crescita Inquinamento Benessere, edited by Musu, I., and Siniscalco, D. Bologna.

Torborg, R.H., 1994, "Capital Budgeting for Environmental Professionals", Pollution Prevention Review, Autumn 1994.

Obu-Asiege, 1992, Okobilanz fur Unternehemen. St Gallen.

United Nations, 1991, International Accounting and Reporting Issues: 1990 Review, New York: United Nations.

United Nations, 1992, International Accounting and Reporting Issues: 1991 Review. New York: United Nations.

United Nations, 1993, Environmental Accounting and Reporting Issues: 1992 Review. New York: United Nations.

United Nations, 1992, Handbook of National Accounting. Integrated Environmental and Economic Accounting. New York: United Nations.

United Nations, 1993, System of National Accounts. New York: United Nations.

US Department of Commerce (Bureau of the Census), 1992, Pollution Abatement Cost and Expenditures. Washington: US Department of Commerce.

---

[1] During the thirteenth session of ISAR, Mr. Matteo Bartolomeo of FEEM made a presentation of the environmental balance sheet methodology. This chapter is a reproduction of the case study as presented. The opinions expressed within this chapter are those of Mr. Bartolomeo and should not be considered to represent those of UNCTAD.

Chapter VI

THE EUROPEAN COMMUNITY'S ECO-MANAGEMENT AND AUDIT SCHEME

Summary

When the European Commission (EC) first drafted its proposals on the scheme in December 1990, the draft was unique in Europe, and possibly the world, as being the first piece of environmental legislation to define a system which included both environmental auditing and public reporting of environmental information.

On July 13 1993, the European Union's Eco-Management and Audit Scheme (EMAS) became effective. The scheme enables industrial companies to meet, on a voluntary basis, environmental protection goals which they have established for themselves, yet compliance with these goals is audited externally. The logic of the scheme is to use market forces to stimulate continual environmental performance improvements within industry by committing corporations to:

(a)   establish and implement environmental policies, programmes and management systems;

(b)   periodically evaluate, in a systematic and objective way, the performance of site elements; and,

(c)   provide environmental performance information to the public.

Companies which participate in the scheme must provide full details to the public of their environmental performance within agreed norms of commercial confidentiality.

This approach represents a fundamentally new strategy towards environmental regulation. Although the regulation's jurisdiction is limited to member countries of the EU, there are no reasons to prevent companies in other countries from using the scheme's elements as a way to manage and improve their environmental performance.

This chapter is a briefing paper which was prepared by the EC on EMAS and was presented at the thirteenth session of ISAR by Mr. Peter Wilson, a representative of the EC.[1]

INTRODUCTION

The overall objective of the European Union (EU), as defined by the Maastricht Treaty, is to promote a harmonious and balanced development of economic activities, sustainable and non-inflationary growth with due respect to the environment, and to the raise the standard of living and the quality of life of the residents of Member States.

More than 200 legal instruments have been adopted so far in the field of environmental protection. In some Member States instruments such as regulations and directives are the basis for most of the environmental legislation in force. The purpose of these instruments is to:

(a)   ensure that harmonised regulatory requirements are in place in order to prevent obstacles and distortions to the internal market, while at the same time, achieve high levels of environmental protection in all member states;

(b) ensure that the contribution to European economic development by the EU includes qualitative as well as quantitative factors; and,

(c) contribute effectively to international debate and activities on environmental matters with coherent policies.

To date the EU's activities in this area have been mainly regulatory. However, in the light of experience gained during the twenty years of EU environmental policy, it appears that the traditional "command and control" approach is no longer sufficient. In spite of all the directives and regulations adopted by the EC, and the international and national promulgation and pronouncements in this field, the quality of the environment continues to deteriorate in the EU and world wide.

The goal of sustainable development, which is now integrated into the EU objectives, calls for the use of a wider range of tools for environmental policy. A new approach is needed and it should be based on different principles of action. The 5th EU Environmental Action Plan recognises the current predicament and clearly indicates that environmental responsibility should be shared between authorities, industry, consumers and the general public.

Moreover, in order to be effective, environmental concerns must be integrated into the formulation of all policies by the various participants in this process. This principle does not only apply to the public authorities. Companies need to integrate environmental considerations into their own policy, strategy and management systems over and above the minimum regulatory requirements.

On 29 June 1993 the European Council adopted a proposal from the EC which allows voluntary participation, by companies in the industrial sector, in a community Eco-management and audit scheme. The regulation establishes a voluntary environmental management scheme, based on harmonised concepts and principles throughout the EU, which may be adopted by

companies which operate in the EU's industrial sector. The regulation is now in force however, it will only be open for participation by companies from April 1995.

I. AN OVERVIEW OF THE EC ECO-MANAGEMENT AND AUDIT SCHEME

A. Objectives

The overall objective of the scheme is to promote continuous improvements of environmental performance by industrial corporations by committing sites to evaluate and improve their environmental performance and to provide relevant information to the public. The scheme does not replace current EU or national environmental legislation or technical standards, nor does it, in any way, remove a company's responsibility to fulfil all its legal obligations under such legislation or standards.

B. Participation

Participation in the scheme is site based and is open to companies which undertake industrial activities. However, Member States have the opportunity to extend the scheme's provisions, on an experimental basis, to other sectors. Upon registration, companies are required to formulate and adopt an environmental policy which contains commitments both to comply with all relevant environmental legislation and also to achieve continuous improvements in environmental performance.

At the site, an initial environmental review is undertaken. In the light of this review, an environmental programme and environmental management system is established for the site. Subsequent environmental audits of the site, in which all activities are audited, must be conducted within an audit cycle of no longer than 3 years. Based on the audit findings and the environmental objectives set, the environmental programme is revised in order to achieve the set objectives. On completion of the initial environmental review and subsequent audits or audit cycles, a public environmental statement is produced.

C. The environmental statement

The public environmental statement and its validation by accredited environmental verifiers is fundamental to the EMAS. The regulation specifies the following requirements for the environmental statement:

(a)   the statement shall be prepared after the initial review and the completion of each subsequent audit or audit cycle for every site which participates in the scheme (Art 5 1.);

(b)   the environmental statement shall be designed for the public and written in a concise and comprehensible form (Art 5 2.);

(c)   the environmental statement shall include in particular the following (Art 5 3.):

(i)   a description of the company's activities at the site considered;

(ii)   an assessment of all the significant environmental issues of relevance to the activities concerned;

(iii)   a summary of the figures on pollutant emissions, waste generation, consumption of raw materials, energy and water, noise and other significant environmental aspects, as appropriate;

(iv)   other factors regarding environmental performance;

(v)   a presentation of the company's environmental policy, programme and management system implemented at the site concerned;

(vi)   the deadline for submission of the next statement; and,

(vii)   the name of the accredited environmental verifier.

D. Registration

Site registration occurs once the competent body, as designated by the Member State, receives a validated environmental statement, any applicable registration fee which is levied, and is satisfied that the site meets the regulation's requirements, which includes compliance with all relevant environmental legalisation. Each year the lists of registered sites from the 15 Member States will be communicated to the EC and a complete list will be published in the Official Journal of the European Communities.

E. Statement of participation

A graphic symbol linked to statements of participation which lists sites, within a given company, which are registered to the scheme. Companies may use the symbol to publicise and promote their involvement in the scheme however, it may not be used in product advertising or on products or their packaging. The graphic symbol may not be used on its own.

F. Annexes of the regulation

The annexes of the regulation provide supplementary details on the requirements which concern environmental policies, programmes, management systems and environmental auditing (Annexes I and II) as well as requirements which concern the accreditation of environmental verifiers and their function (Annex III). Annex IV details the four different statements of participation that sites may use to indicate their successful inclusion in the scheme and the graphic symbol and Annex V outlines information that a company must present to the competent body. The annexes maybe modified in the light of experience gained during the operation of the scheme.

G. Revision

EMAS will be revised after 5 years of operation at which time, in the light of experience gained, the EC can propose appropriate amendments to the Council. Amendments could include the possible introduction of a logo.

II.   FEATURES OF EMAS

A. Mechanism for standardisation

Companies which have implemented and have been certified as having met national, European or international standards in order to meet certain aspects of the scheme, such as its requirement for an environmental management system, will be deemed to have met those parts of the regulation. This is on the basis that the standards used fulfil two conditions. Firstly, the standards must be recognised by the EC and secondly, the standards must be certified by a body whose accreditation is recognised by the Member State where the site is located.

Decisions will be required to cover both the standard and the relevant certification system.

B. The EMAS verifier

Accredited environmental verifiers have two clear roles. These are:

(a)   firstly, to check that the elements of EMAS i.e. the environmental policy, management system, programme, review and audit are in place, operational and conducted in accordance with the appropriate specifications contained within the annexes; and,

(b)   secondly, in addition to an examination of the reliability and coverage of the information in the environmental statement, the verifier must validate that information.

Within the regulation it is foreseen that the verifier can either be an organisation or, with limited accreditation scope, an individual. The verifier does not in any way replace the Member States' environmental regulatory authorities. Verifiers can operate in any Member State but they must notify, and are supervised by, the Member State system in which they perform their validation activities. Details on the accreditation of environmental verifiers and their function are outlined in the Annex III of the regulation.

Figure VI.1 *Key Stages in The Verification process* [2]

Examine Documentation

- environmental review
- internal audit methodologies and programmes
- procedures manuals
- management and audit reports
- performance data
- policy and description of environmental management system

then

Visit Site

- Include meetings with personnel, in particular internal auditors and those responsible for operational controls

then

Prepare a Verification report

- cases of non-compliance with EMAS requirements
- technical defects in procedures
- amendments/additions required to the statements

then

Liaison and Verification

- liaison on solutions to issues raised (prior to registration)
- verifiers statement and signature

C. Special activities with SMEs

Efforts have been made within the regulation to introduce provisions to assist small and medium sized enterprises (SMEs) involvement in the scheme. Member States may promote participation by companies of all sizes, but in particular SMEs, by establishing technical assistance which would help companies meet the regulation's requirements, for example to establish environmental policies, programmes and management systems. The EC will also consider measures, especially training, structural and technical support, to achieve greater participation of SMEs.

III. THE FUTURE OF EMAS

A. Within the commission

The EC has a number of activities which are a follow up to the regulation which has been introduced. In particular the following activities are under way:

(a)   continuation of work to develop decisions on equivalence of the national standards;

(b)   formulation of proposals to the Council of Ministers on BS7750;

(c)   provision of guidance on audit periodicity;

(d)   provision of guidance on criteria for accreditation of verifiers;

(e)   formulation of guidelines for the activities of verifiers;

(f)   publication of the Guide to the Eco-Management and Audit Scheme; and

(g)   a workshop on pilot experiences of organizations in the process of implementing EMAS.

B. Activities within the member states

Member States of the EU have began to implement the scheme and the following are ways in which this is being done:

(a)   designation of competent authorities (where this has not already been done);

(b)   establishment of registers of verifiers and of registered sites;

(c)   designation of accreditation bodies and accreditation of verifiers; and,

(d)   promotion of the scheme and to increase awareness thereof.

It is intended that all the necessary components of the scheme will be in place to allow companies to be verified and registered in the scheme by April 1995 as originally envisaged. It should be remembered that these are the first concrete steps in this area and a great deal can be learnt from the actual initial implementation process. Close cooperation among all the parties involved will be essential.

The basic idea behind these initiatives is that environmental care is not just a matter of technical regulations, and the attainment of standards and inspections. In order for industry to contribute to sustainable development, the full motivation and involvement of a company's management and personnel, at all levels, is required. The involvement of the public through increased awareness and participation, and the dissemination of information are also crucial elements.

An open and transparent attitude by industry, a willingness to participate in dialogue by public authorities and an increased level of awareness and maturity on behalf of the public, are all preconditions for a new partnership in the field of the environment by all the participants. The trend towards a greener culture in the business community must be actively encouraged. With EMAS, the EU and its Member States intend to be at the forefront of such processes.

Notes

[1]   The presentation was based upon Council Regulation EEC/1836/93 of 29 June 1993. The opinions expressed within this chapter represent those of the author and should not be interepreted to be those of UNCTAD.

[2]   Extracted from the Eco-Management and Audit Scheme introductory guide.

Chapter VII

SUSTAINABLE FORESTRY OPERATIONS AND ACCOUNTANCY

Summary

The UNCTAD publication *Accounting for Sustainable Forestry Management: A case study* [1] was presented to the thirteenth session of ISAR. This study resulted from extensive research primarily undertaken by Mr. Daniel Rubenstein of the Auditor General's Office, Canada. The research involved the cooperation of many individuals, corporations and academic institutions to which UNCTAD expresses its gratitude. It was guided by a multidisciplinary team of experts consisting of engineers, chemists, foresters, mill operators, and environmentalists, as well as accountants, who met periodically to guide and contribute to the development of the case study. This chapter contains: the report of the final workshop which reviewed the case study before it was published; the presentation made by Mr. Rubenstein at ISAR's thirteenth session of his experiences while researching the case study; and a summary of the comments made about this publication by the ISAR experts present at the session.

I. THE REPORT OF THE FINAL WORKSHOP ON THE CASE STUDY

As part of the study, a series of workshops were convened to review the research methodology to be used for the case study and to identify practical issues which should be addressed. Prior to completion of the report, on 13 and 14 May 1993 a final ad hoc workshop of experts reviewed the results of the research and provided comments to the principal researcher.

A. The participants

The following list demonstrates the diverse backgrounds of the participants. The previous workshops were attended by some of the experts listed below and various other experts, all of whom were selected based on their expertise in the respective field of the case study under review.

a) Ms. Jane B. Adams
   The United States
   Financial Accounting
   Standards Board
   U.S.A.
b) Mr. Michael Barkwell
   Petro Canada
   Canada
c) Professor Jan Bebbington
   Department of Accountancy and
   Business Finance
   The University of Dundee
   United Kingdom
d) Professor Soren Bergstrom
   Stockholm University
   Sweden
e) Ms. Kristin Dawkins
   The Institute for Agriculture and
   Trade Policy
   U.S.A.

f) Ms. Suzi Evalenko
    Trade Policy
    American Forest and Paper
       Association
    U.S.A.
g) Mr. Frederick Gill
    American Institute of Certified Public
       Accountants
    U.S.A.
h) Mr. James Goodfellow
    Deloitte & Touche Chartered
       Accountants
    Canada
i) Mr. John Heissenbuttel
    Private Forestry
    American Forest & Paper Association
    U.S.A.
j) Mr. Tom Leonard
    Accounting Projects
    General Electric Corporation
    U.S.A.
k) Mr. David Moore
    The Canadian Institute of Chartered
       Accountants
    Canada
l) Mr. John Niamon
    Investor Responsibility Research
       Centre Inc.
    U.S.A.
m) Mr. Tej Prekash
    Government of India
    India
n) Mr. Tony Rotherham
    Canadian Pulp and Paper Association
    Canada
o) Mr. Daniel Rubenstein
    Office of the Auditor General of
       Canada
    Canada
p) Prof. Eammon Walsh
    Stern School of Business New
       York University
    U.S.A.

B. The issues raised

Six broad issues were raised during the workshop, as described below. The case study report was subsequently modified based upon these comments.

(a) The balance between economic sustainability and ecological sustainability was analyzed. The management of Kirkland Forest Products Ltd., the fictitious forestry enterprise in the case, was allowed to define for itself the term "sustainable development" and then to hypothesize changes to commercial operations in order to meet that definition. The management proposed three scenarios to contrast the implications of any change:

(i)   a "business as usual" approach;
(ii) modifying the operations of the company in order to meet management's definition of sustainable development; and,
(iii) taking a "deep green" approach whereby environmental considerations are of the utmost importance and drive all decisions by management.

It became apparent that the enterprise's choice of how its operations could be modified was significantly influenced by the fact that it had recently constructed a pulp processing plant which required a certain throughput volume of production in order for the plant to remain economically viable. Therefore, the participants felt that the enterprise's definition of sustainability was considerably influenced by economic considerations rather than purely ecological factors. Hence, the definition was really an "industrial" definition.

(b) The valuation of wood and non-wood assets was considered. In order to measure sustainability the traditional criteria for assets must be expanded to comply with the wider definition of ecological stewardship which is required under the concept of sustainability. The most difficult task was to value these newly defined assets many of which are intangible. A number of methods can be used to value enhanced wood assets; however the group thought it might be safer to record just enhanced physical volumes. The group rejected giving infinite values to non-wood assets but urged that some inherent, non- monetary value or some relative scale be applied to measure changes or improvements to these intangibles.

(c) The relationship between financial accounting and sustainable development was discussed. Some of the participants questioned whether financial accounting could contribute to the subject. If so, the objectives and standards of sustainability would have to be more precise to allow the rigid principles of financial accounting to be applied effectively. Practising accountants emphasized that companies were already under pressure to expand the domain of accountability, and the current trend to disclose environmental information is in response to this pressure. Accounting records are the primary source of this information and report upon the consumption and generation of financial resources. To date, financial capital only has been recorded, but natural capital, human capital and consumer capital could also be taken into account.

It was suggested that industry associations define "sustainability" as applicable to themselves. Subsequently, companies could define their objectives, set benchmarks, and make a commitment towards attainment of these goals. Accountants can help measure the progress.

(d) The effect of environmental performance information upon capital markets was debated. Some of the participants felt that the public would like to see capital markets operate in a more responsible manner towards the environment. This would entail acceptance of the fact that sustainable corporations are unable to offer the same level of financial returns *vis-a-vis* their competitors whose *modus operandi* maintains the status quo. An accounting system discloses information to the capital markets. Accounting is an integral part of disclosure and there is a need to expand the definition of capital to consider non-monetary as well as monetary values.

Currently, there appears to be serious flaws in the capital markets because environmental responsibilities and liabilities are not truly reported by corporations. Therefore capital decisions are made on the basis of incomplete information. For instance, the cost of the United States' so-called "superfund" cleanup is estimated to be 7 trillion US dollars yet corporations in the aggregate report considerably less liabilities for these obligations.

Investors are usually primarily interested in the internal rate of return on investments and accordingly prefer all appropriate information to

be integrated into the financial statements. Special environmental reports may not always serve the needs of investors.

(e) There is a danger in compressing environmental performance indicators into one number. The case study allowed the management of the enterprise to compute the cost of attaining its definition of sustainable development and an estimate of "sustainable" income. Some of the experts felt that this number was too synthetic and that it might hide too much. However, others felt that there was merit in focusing upon the additional expenditures which are required in order to minimize the environmental impact of commercial activities.

(f) Some of the experts questioned whether accounting for sustainable development is an internal management tool or more appropriately a financial reporting method. This question can be applied to the case study. It was felt that environmental accounting, as portrayed in the case study, has not yet been sufficiently refined to enable the information to be reported to the shareholders; however, the information could assist managers over time to put into effect the concept of sustainable development. Management needs to set policies to implement and monitor before such information is reported publicly. One expert suggested that companies must improve their environmental performance before information should be disclosed to the public.

C. Suggestions for future activities

The review workshop concluded with the following recommendations for future work by ISAR in the field of environmental accounting:

(a) Perform a survey, by industry, of the various approaches to environmental accounting which have been developed and identify practical applications and approaches which have evolved in each industry studied. Hence, it will be possible to determine which areas of environmental accounting have been adequately researched and in which industries methodologies have been formulated and identify aspects of environmental accounting which have evolved the least;

(b)   Develop a select number of environmental performance indicators which are able to measure improvements in environmental impact and the related financial implications, and relate these improvements to the original targets that were set by enterprises;

(c)   Undertake further case studies to re-affirm the applicability of the approach that was taken in this case study;

(d)   Develop a link between environmental accounting at the macro level, that is the national income level, and at the enterprise level; and finally,

(e)   Investigate incentives and disincentives for sustainable development in the international business community. This would entail inquiry as to the perceptions of Chief Executive Officers about the estimated costs and benefits involved in the adoption of sustainable development programmes.

II. A DESCRIPTION OF THE RESEARCH BEHIND THE CASE STUDY

The principal researcher described the measures undertaken to develop the case study and his remarks added considerable credibility to the contents of the publication. The following is a brief summary of the points covered during his presentation.

A. The financial estimates

Of particular concern to some experts was the "soft nature" of the figures which are used throughout the case study of Kirkland Forest Products Ltd. (Kirkland). The publication notes that the figures are fictitious; however, Mr. Rubenstein assured the experts that the figures are based upon estimates that are as realistic as possible. The case study was modeled upon an existing corporation which operates in an area of Canada not dissimilar from that in which Kirkland is hypothetically located. Hence the cost accounting throughout the case study is as realistic as possible. Mr. Rubenstein had agreed to provide confidentiality about information which had been divulged to him by company representatives.

As an example, attention was drawn to table III, "Range of sustainability for forest operations". The figures of $4.8 million and $7.2 million are estimates, but are as good an estimate as is ever made for long term financial plans. See below for details of the approach taken to make the estimates.

B. Tolerable limits of environmental impact

The concept of sustainable development has been debated extensively and many ideas have evolved; however there is no clear directive for the actual implementation of these theories. When the case study was conceived one of the first steps was to determine the bounds of environmental impact which could be tolerated, and it is within these bounds that the company has self-defined sustainability from a commercial viewpoint. It was suggested that the relationship between business and sustainable development is one of compromise, and the relationship is essentially a very precarious balance between environmental care whilst maintaining business viability. Furthermore, within society different groups have their individual objectives and interests. Environmentalists are in favour of complete bio-diversity which would entail extensive protection measures. Simultaneously, there are many people employed in the forestry areas whose livelihood is dependent upon logging, milling and pulping activities. Clearly, compromises must be made in order to reach mutually agreeable solutions.

In fact, the case study develops three different scenarios for consideration, each with various degrees of environmental impact. The first option is a situation where minimum environmental protection legislation is met. The second possibility is intended to result in a level of environmental impact which is more stringent than minimum compliance. The corporation would be making a selective response to concerns expressed by the general public and accordingly would provide greater protection to the forest -- care for the habitat, which the forest provides for the wildlife and fisheries, and the forest's aesthetic beauty. Nevertheless, management of the forest is essentially based upon economic principles. The third scenario considers a situation in which the volume of production and the technology employed in the production process are determined by ecological and not economic considerations. Mr. Rubenstein stated that this third scenario is the crux of a sustainable relationship between industrial activity and environmental management.

C. The approach taken

Mr. Rubenstein felt that he had gained a very comprehensive insight into the industry during his numerous interviews of foresters, timber fellers, saw mill operators, engineers, industrial chemists, environmentalists and other professionals and workers. This insight enabled him to prepare realistic estimates of the numbers in the case study. For instance, estimates had to be made of the cost increases to: change the size of the clear cuts; construct alternate access roads; modify the types of herbicides and bleaching techniques which could be used; and, provide for the extent to which forest areas needed to remain untouched in order to minimize the disturbance to the fauna's habitat. These estimates were all calculated with the advice of experts.

He continued to explain that Kirkland does not practice clear-cutting, and in fact, the corporation has a forest management programme (refer to page 33 of the case study). Furthermore, the term "clear-cut" is often misunderstood. The widely held perception of a clear-cut is to completely remove all of the trees within the harvest area. However, in actual practice the felling of trees is dependent upon the species, topography and various other factors in order to minimize the environmental impact. In Ontario, Canada legislation requires the timber industry to re-plant trees in harvested areas. For additional data, Mr. Rubenstein interviewed representatives of an aboriginal community of indigenous people in Ontario in order to obtain their inputs for the case study, particularly when deriving a working definition of sustainability. This is described within the case study as the "seven generation" test.

D. The case study in detail

Mr. Rubenstein proceeded to explain a number of the tables within the publication. Of

particular concern was table VI which is titled "Price implications of a zero impact pulp mill sustaining non-wood values". This table details the increase in prices for the commodities of pulp and lumber which are required to meet the standards set under sustainability options 2 and 3 as defined in the case study. Accountants calculated the cost implications of the policy options and the sales price increases needed to cover the increased costs. This information was combined with capital budget requirements to demonstrate to Kirkland's management the differences in the rate of return on investment of the three scenarios which are discussed throughout the case study (refer to table VII).

To quantify in monetary terms the changes to the return on investment is relatively straight forward; however Mr. Rubenstein expressed the opinion that it was more difficult to identify for management the incentives to making this transition, particularly because in many situations the benefits are invisible over the short term and difficult to quantify *vis-a-vis* the cost increases which are immediate and considerable. Chapter II of this publication debates this issue.

Mr. Rubenstein acknowledged that virtually all of the numbers within the case study are estimates, are subjective, and may be challenged. Nevertheless, he thought that accountants had a responsibility towards their managements and clients to at least make an attempt at quantifying the costs of enhanced environmental impact rather than potentially misleading them by remaining silent on the topic. He concluded that "it is better to be approximately right than precisely wrong".

E. Repeating the case study

ISAR experts were urged to undertake similar studies in their home countries. This case study is essentially a model which sets out a methodology as to how commercial principles can be applied to sustainability concepts and how commercial practices should be modified in order to determine the cost and benefits, the sacrifices and the implications of a move towards sustainability.

In order to build up subsequent case studies, accountants must work in multi-disciplinary teams in order to obtain expert opinions from other professionals and skilled individuals who are able to contribute their expert knowledge to ensure that projections are as realistic as possible. Clearly, the results are estimates; however, it must be remembered that a great number of business decisions are based upon estimates. For instance, the valuation of fixed assets, cost budgeting and investment analyses are all based upon projections, and so the methodology applied in the case study is a legitimate exercise in accountancy. Cost accounting expertise is invaluable in the development of different scenarios which depict various levels of environmental impact. This information can assist stockholders, consumers and management to choose between the reduction of profits and environmental damage because the cost implications of policy options are quantified.

III. QUESTIONS AND COMMENTS FROM THE ISAR EXPERTS

Professor Robert Gray, a consultant to the Group, commented that the case study is an important example of the kinds of experimentation on environmental accounting and reporting matters that need to be performed. The case study illustrates that it is not possible to come up with a quick solution to accounting for the environment. However, accounting can help measure the extent of the problem of where an organization is now and what it needs to do to attain sustainability.

At this stage of development, attempts are not being made to replace conventional financial accounting but it is necessary to be clear on the objectives that are trying to be addressed. Accounting for sustainability is very demanding and requires the talents of everyone. Also, some believe that it is too premature to report on environmental matters because there is a need for more experimentation. Some questions raised are: What is the accounting entity? What is its purpose? Whose sustainability are we looking for?

Although conventional accounting is difficult enough, accountants have an obligation to serve the interests of the public and survival of the planet is part of that interest.

The representative from the Netherlands asked about the distinction between economic and ecological sustainability. In the case study it was pointed that the management chose between economic sustainability and not ecological sustainability. He wondered whether this was a fair comparison of the two conceptual models since the notion is to achieve a proper balance between economic and ecological sustainability.

Mr. Rubenstein agreed that there is a dynamic tension that exists within the management of an enterprise, such as the one included in the case study, between establishing economic and ecological criteria in trying to define what is a sustainably balanced development.

The representative from Gabon asked Mr. Rubenstein for a tangible definition of the term, "sustainable development" as it was used in preparing the case study. Mr. Rubenstein responded that there is no simple definition because it must be in terms of a given time and place that reflects the unique conditions that exists in each country.

The representative of Germany stated that the case study shows that accounting in the traditional form is helpful for measuring and calculating costs and prices and it is able to make clear that we cannot protect the environment at no cost to the public. This must be paid for. Therefore it is very important to develop financial accounting methods for all interested persons that provide sufficient information on the costs of environmental protection. Accounting is able to make clear that rehabilitation and recultivation of a country's natural capital are costs of the products derived therefrom and should be included in pricing those products to consumers. Prices should cover all costs.

The representative stressed that companies operating in countries which require environmental restoration have to bear costs which may make those companies' products uncompetitive in international markets where there are producers from other countries that do not have to incur such costs. In order to encourage more responsibility for natural resources, governments around the world must jointly regulate greater accountability for natural resources costs to ensure that corporations in an individual country act more responsibly towards the environment without being penalized by market forces.

Furthermore, investors and creditors provide capital to companies that are profitable, and if companies that have to include the costs of rehabilitation and restoration in their costs become unprofitable they will not be able to attract sufficient capital with which to operate. This will encourage the expansion of operations in those countries where companies do not have to bear environmental restoration costs and this will lead to all citizens of the world suffering from the resultant damage to the environment.

ISAR should discuss at a later date how governments should account for natural capital and develop recommendations to governments for use in making the political decisions that are required for environmental protection.

Mr. Rubenstein concurred with the views expressed by the representative of Germany. Clearly, the political aspects of a solution to the problem on accounting for environmental costs had to be resolved at the international level, preferable through the United Nations. He suggested that standards be developed which would allow consumers from anywhere in the world to know, on a uniform basis, the real cost of products. These costs would not necessarily have to be reflected in the price of products, but at least packaging or other documentation should be used to keep consumers informed.

The Secretariat intervened at this point and informed the Group that in February 1995 there was important meeting in Geneva where this issue was discussed. The meeting was the High-Level Session, "Environment and Trade: Perspectives of Developing Countries", which was sponsored by UNCTAD and UNEP, the United Nations Development Programme. Several participants in that meeting pointed to the need for market prices of products based on their true costs of products. There was broad agreement at the meeting on the need for progress to be made on the so-called internalization of environment externalities as a means to adjust environmental underpricing. Also,

there was strong agreement among both developing and developed country participants in the meeting that trade represents a fundamental means with which to build sustainable development, and therefore it is vitally important that cost internalization takes place for product pricing purposes. However, restrictive trade policies based on environmental concerns are unacceptable, particularly to developing countries, and therefore policies on the internalization of costs involve both political and economic issues.

Professor Gray remarked that the problem of sustainability which the world is facing is extremely complex and he is concerned that insufficient concrete, practical initiatives to actually implement the notion of sustainability had occurred. He was alarmed that the only real progress in terms of environmental impact that has been made is a reduction in the acceleration of environmental degradation in the world. Economic, market and accountancy theories can only achieve so much, especially when challenged by human nature and natural sciences which severely hinder the applicability of these theories in the real world. Nevertheless, everyone has a moral obligation to make a personal contribution in the quest for solutions to this global challenge.

Next, the representative of Costa Rica agreed that solutions to the problems of the environment depend on political decisions. Costa Rica is in the process of implementing Agenda 21 and for the current government of her country sustainable development is a central theme. She felt that sustainable development doesn't mean conservation but the rational use of natural resources. The experience in Costa Rica is that conservation can result in an adverse impact on the economic development of a country. The pressures of human development drives the consumption of resources and to the extent that renewable resources are used, governments are responsible for that renewal. If resources are non-renewable, then substitutes must be found.

She suggested that developing countries may be more inclined to internalize environmental costs than developed countries because they have diverse ecosystems and internalization of costs is an effective way to maintain this diversity. She could not share the views of some of the industrialized countries that the decision to internalize environmental costs should be postponed, because it is the industrialized countries that are the greatest polluters.

Costa Rica will adopt the model of environmentally adjusted national accounts that has been developed by the United Nations Statistical Division as a basis for analysis and discussion of environmental matters. For this activity, the Government has received support from international organizations such as the United Nations Development Programme and the World Bank.

The representative of Italy then tried to establish a general framework for the essence of the dilemma that is now being faced. In principle, accountants only deal with market values so it is challenging to address environmental matters which are not taken into account by the market. The social costs of economic and industrial activities which are not taken into account are highly subjective and depend on the scientific knowledge that humankind has at this stage of world development. With the progress of science these costs change so it is very difficult to address the problem of quantification of environmental costs. But because it is so important a problem, accountants, just as economists and the members of other professions, can not fail to address the problem.

Therefore, he felt that the failure of the private sector to deal with the costs and benefits of the environment is the principal reason for government intervention. Accountants must work with other professionals to overcome this deficiency which is the basis for the survival of mankind. He also feels that the objective should be to internalize externalities. He would support a proposal for the United Nations to organize a group to perform research and develop a framework on environmental accounting.

Next, the representative of the Netherlands asked whether the calculations in the case study had included the enhanced commercial value of the most sustainable option of operating the enterprise (referred to as option 3 in the case study). Mr. Rubenstein replied that any potential increase in value had not been taken into account in the calculations.

The representative from Bulgaria commented that economic information, approximately 80 per cent of which is derived from accounting data, is very often not used in the political decision-making process. As a consequence, politicians may take actions which are contrary to what is contained in the economic information, and the results are almost always negative. In order to solve the problems of the environment, governments must use accounting information as a data base for economic and political solutions.

A representative of Brazil expressed the opinion that the development of the process of environmental accounting calls for the formation of a United Nations study group in view of the challenges of involving various scientific disciplines, as was stated previously by the representative of Italy.

Lastly, the representative of Switzerland commented that he felt that some of the participants believe that the reporting of enterprise performance should be confined to the comparatively narrow aspects of financial accountancy and other means, such as management accounting, which are traditionally used. He was not certain that this is the best approach. His conclusion from the presentations and discussions at this session was that there can be no quick and easy solution to the difficulties posed by environmental accounting. Nevertheless, progress can be made by a pragmatic step-by-step approach. Certain aspects of current accounting methodologies can be improved to deal with environmental matters. There are many options for disclosing more of the real costs and risks to the environment that exist. What is most important is to keep moving ahead or it will not be possible to achieve what needs to be achieved with regard to accounting for environmental matters. The existing financial accounting methods should be continued but should be supplemented with new ideas. For example, the forestry case study methodology should be adapted to other industries, such as the packaging industry, which could facilitate the development of new ideas for finding a solution to the very difficult task of accounting and reporting for environmental matters.

Notes

[1] *Accounting for Sustainable Forestry Management: A case study* (UNCTAD publication, sale number E.94.II.a.17).

Select list of publications of the
UNCTAD Division on Transnational Corporations and Investment

A. Individual studies

International Accounting and Reporting annual issues:
  1989 Review. 152 p. Sales No. E.90.II.A.4.
  1990 Review. 254 p. Sales No. E.91.II.A.3.
  1991 Review. 244 p. Sales No. E.92.II.A.8.
  1992 Review. 306 p. Sales No. E.93.II.A.6.
  1993 Review. 249 p. Sales No. E.94.II.A.16.
  1994 Review. 94 p. Sales No. E.95.II.A.3.

Accounting for Sustainable Forestry Management: A case study. 46 p. Sales No. E.94.II.A.17. $22.

Conclusions on Accounting and Reporting by Transnational Corporations. 47 p. Sales E.94.II.A.9. $25.

Accounting, Valuation and Privatization. 190 p. Sales No. E.94.II.A.3. $25.

Environmental Accounting: Current Issues Abstracts and Bibliography. 86 p. Sales No. E.92.II.A.23.

Small and Medium-sized Transnational Corporations: Executive Summary and Report on the Osaka Conference. 60 p. UNCTAD/DTCI/6. Free-of-charge.

World Investment Report 1994: Transnational Corporations, Employment and the Workplace. 482 p. Sales No. E.94.II.A.14. $45.

World Investment Report 1994: Transnational Corporations, Employment and the Workplace. An Executive Summary. 34 p. Free-of-charge.

World Investment Directory. Volume IV: Latin America and the Caribbean. 478 p. Sales No. E.94.II.A.10. $65.

Liberalizing International Transactions in Services: A Handbook. 182 p. Sales No. E.94.II.A.11. $45. (Joint publication with the World Bank.)

Environmental Management in Transnational Corporations: Report on the Benchmark Corporate Environment Survey. 266 p. Sales No. E.94.II.A.2. $29.95.

Management Consulting: A Survey of the Industry and Its Largest Firms. 100 p. Sales No. E.93.II.A.17. $25.

Transnational Corporations: A Selective Bibliography, 1991-1992. 736 p. Sales No. E.93.II.A.16. $75.

Small and Medium-sized Transnational Corporations: Role, Impact and Policy Implications. 242 p. Sales No. E.93.II.A.15. $35.

World Investment Report 1993: Transnational Corporations and Integrated International Production. 290 p. Sales No. E.93.II.A.14. $45.

World Investment Report 1993: Transnational Corporations and Integrated International Production. An Executive Summary. 31 p. ST/CTC/159. Free-of-charge.

Foreign Investment and Trade Linkages in Developing Countries. 108 p. Sales No. E.93.II.A.12. $18.

World Investment Directory 1992. Volume III: Developed Countries. 532 p. Sales No. E.93.II.A.9. $75.

Transnational Corporations from Developing Countries: Impact on Their Home Countries. 116 p. Sales No. E.93.II.A.8. $15.

Debt-Equity Swaps and Development. 150 p. Sales No. E.93.II.A.7. $35.

From the Common Market to EC 92: Regional Economic Integration in the European Community and Transnational Corporations. 134 p. Sales No. E.93.II.A.2. $25.

World Investment Directory 1992. Volume II: Central and Eastern Europe. 432 p. Sales No. E.93.II.A.1. $65. (Joint publication with ECE.)

World Investment Directory 1992. Volume I: Asia and the Pacific. 356 p. Sales No. E.92.II.A.11. $65.

B. Serial publications

UNCTC Current Studies, Series A

No. 18.   *Foreign Direct Investment and Industrial Restructuring in Mexico*. 114 p. Sales No. E.92.II.A.9. $12.

No. 19.   *New Issues in the Uruguay Round of Multilateral Trade Negotiations*. 52 p. Sales No. E.90.II.A.15. $12.50.

No. 20.   *Foreign Direct Investment, Debt and Home Country Policies*. 50 p. Sales No. E.90.II.A.16. $12.

No. 22.   *Transnational Banks and the External Indebtedness of Developing Countries: Impact of Regulatory Changes*. 48 p. Sales No. E.92.II.A.10. $12.

No. 23.   *The Transnationalization of Service Industries: An Empirical Analysis of the Determinants of Foreign Direct Investment by Transnational Service Corporations*. 62 p. Sales No. E.93.II.A.3. $15.00.

No. 24.   *Intellectual Property Rights and Foreign Direct Investment*. 108 p. Sales No. E.93.II.A.10. $20.

No. 25.   *International Tradability in Insurance Services*. 54 p. Sales No. E.93.II.A.11. $20.

No. 26.   *Explaining and Forecasting Regional Flows of Foreign Direct Investment*. 58 p. Sales No. E.94.II.A.5. $25.

No. 27.   *The Tradability of Banking Services: Impact and Implications*. 195 p. Sales No. E.94.II.A.12. $50.

No. 28.   *Foreign Direct Investment in Africa*. 119 p. Sales No. E.95.II.A.6. $25

The United Nations Library on Transnational Corporations. (Published by Routledge on behalf of the United Nations.)

Set A (Boxed set of 4 volumes. ISBN 0-415-08554-3. £350):

Volume One: *The Theory of Transnational Corporations*. 464 p.

Volume Two: *Transnational Corporations: A Historical Perspective*. 464 p.

Volume Three: *Transnational Corporations and Economic Development*. 448 p.

Volume Four: *Transnational Corporations and Business Strategy*. 416 p.

Set B (Boxed set of 4 volumes. ISBN 0-415-08555-1. £350):

Volume Five: *International Financial Management*. 400 p.

Volume Six: *Organization of Transnational Corporations*. 400 p.

Volume Seven: *Governments and Transnational Corporations*. 352 p.

Volume Eight: *Transnational Corporations and International Trade and Payments*. 320 p.

Set C (Boxed set of 4 volumes. ISBN 0-415-08556-X. £350):

Volume Nine: *Transnational Corporations and Regional Economic Integration*. 331 p.

Volume Ten: *Transnational Corporations and the Exploitation of Natural Resources*. 397 p.

Volume Eleven: *Transnational Corporations and Industrialization*. 425 p.

Volume Twelve: *Transnational Corporations in Services*. 437 p.

Set D (Boxed set of 4 volumes. ISBN 0-415-08557-8. £350):

Volume Thirteen: *Cooperative Forms of Transnational Corporation Activity*. 419 p.

Volume Fourteen: *Transnational Corporations: Transfer Pricing and Taxation*. 330 p.

Volume Fifteen: *Transnational Corporations: Market Structure and Industrial Performance*. 383 p.

Volume Sixteen: *Transnational Corporations and Human Resources*. 429 p.

Set E (Boxed set of 4 volumes. ISBN 0-415-08558-6. £350):

Volume Seventeen: *Transnational Corporations and Innovatory Activities*. 447 p.

Volume Eighteen: *Transnational Corporations and Technology Transfer to Developing Countries*. 486 p.

Volume Nineteen: *Transnational Corporations and National Law*. 322 p.

Volume Twenty: *Transnational Corporations: The International Legal Framework*. 545 p.

Transnational Corporations (formerly *The CTC Reporter*).

Published three times a year. Annual subscription price: $35; individual issues $15.

Transnationals, a quarterly newsletter, is available free of charge.

United Nations publications may be obtained from bookstores and distributors throughout the world. Please consult your bookstore or write to:

United Nations Publications

Sales Section	OR	Sales Section
Room DC2-0853		United Nations Office at Geneva
United Nations Secretariat		Palais des Nations
New York, N.Y. 10017		CH-1211 Geneva 10
U.S.A.		Switzerland
Tel: (1-212) 963-8302 or (800) 253-9646		Tel: (41-22) 917-1234
Fax: (1-212) 963-3489		Fax: (41-22) 917-0123

All prices are quoted in United States dollars.

For further information on the work of the Transnational Corporations and Investment Division, UNCTAD, please address inquiries to:

United Nations Conference on Trade and Development
Division on Transnational Corporations and Investment
Palais des Nations, Room E-8006
CH-1211 Geneva 10
Switzerland

Telephone: (41-22) 907-5707
Telefax: (41-22) 907-0194

QUESTIONNAIRE

International Accounting And Reporting Issues:
1995 Review
(UNCTAD/DTCI/25)

In order to improve the quality and relevance of the work of the United Nations Conference on Trade and Development, it would be useful to receive the views of the readers on this and other similar publications. It would therefore be greatly appreciated if you could complete the following questionnaire and return it to:

Readership Survey
United Nations Conference on Trade and Development
Building E, Room E-9008
Palais des Nations
CH-1211 Geneva 10
Switzerland

1. Name and address of respondent (optional):

   _____

   _____

   _____

2. Which of the following best describes your area of work?

Government	☐	Public enterprise	☐
Private enterprise	☐	Academic or research institution	☐
International organization	☐	Media	☐
Non-profit organization	☐	Other (specify)	☐

3.   In which country do you work? _____

4.   What is your assessment of the contents of this publication?

Excellent   ☐   Adequate   ☐

Good   ☐   Poor   ☐

5.   How useful is this publication to your work

Very Useful   ☐   Of some use   ☐   Irrelevant   ☐

6.   Please indicate the three things you liked best about this publication:

_____

_____

_____

7.   Please indicate the three things you liked least about this publication:

_____

_____

_____

8.   If you have read more than the present DTCI publication, what is your overall assessment of them?

Consistently good   ☐   Usually good but with some exceptions   ☐

Generally mediocre   ☐   Poor   ☐

9.   On the average, how useful are these publications to you in your work?

Very Useful ☐   Of some use ☐   Irrelevant ☐

10.   Are you a regular recipient of *Transnational Corporations* (formerly *The CTC Reporter*), the Division's tri-annual publication which reports on the Division's and related work?

Yes ☐   No ☐

If not, please check here if you would like to receive a sample
copy sent to the name and address you have given above ☐